Universitext

Springer
*Berlin
Heidelberg
New York
Hong Kong
London
Milan
Paris
Tokyo*

Fred Espen Benth

Option Theory with Stochastic Analysis

An Introduction to
Mathematical Finance

 Springer

Fred Espen Benth

Centre of Mathematics for Applications
University of Oslo
Department of Mathematics
P.O. Box 1053 Blindern
0316 Oslo
Norway
e-mail: fredb@math.uio.no

Title of the original Norwegian edition:
Matematisk Finans ©Universitetsforlaget AS, Oslo, 2002

This book has been funded by NORLA-Norwegian Literature Abroad, Fiction and Non-fiction

Cataloging-in-Publication Data applied for
A catalog record for this book is available from the Library of Congress.
Bibliographic information published by Die Deutsche Bibliothek
Die Deutsche Bibliothek lists this publication in the Deutsche Nationalbibliografie; detailed bibliographic data is available in the Internet at http://dnb.ddb.de

Mathematics Subject Classification (2000): 91B28, 60H30, 65C05, 60G35

ISBN 3-540-40502-X Springer-Verlag Berlin Heidelberg New York

This work is subject to copyright. All rights are reserved, whether the whole or part of the material is concerned, specifically the rights of translation, reprinting, reuse of illustrations, recitation, broadcasting, reproduction on microfilm or in any other way, and storage in data banks. Duplication of this publication or parts thereof is permitted only under the provisions of the German Copyright Law of September 9, 1965, in its current version, and permission for use must always be obtained from Springer-Verlag. Violations are liable for prosecution under the German Copyright Law.

Springer-Verlag is a part of Springer Science+Business Media

springeronline.com

© Springer-Verlag Berlin Heidelberg 2004
Printed in Germany

The use of general descriptive names, registered names, trademarks, etc. in this publication does not imply, even in the absence of a specific statement, that such names are exempt from the relevant protective laws and regulations and therefore free for general use.

Cover design: *design & production* GmbH, Heidelberg
Typeset by the author using a Springer LATEX macro package
Printed on acid-free paper   41/3142db- 5 4 3 2 1 0

To my wife Jūratė

Preface

Since 1972 and the appearance of the famous Black & Scholes option pricing formula, derivatives have become an integrated part of everyday life in the financial industry. Options and derivatives are tools to control risk exposure, and used in the strategies of investors speculating in markets like fixed-income, stocks, currencies, commodities and energy.

A combination of mathematical and economical reasoning is used to find the price of a derivatives contract. This book gives an introduction to the theory of *mathematical finance*, which is the modern approach to analyse options and derivatives. Roughly speaking, we can divide mathematical finance into three main directions. In *stochastic finance* the purpose is to use economic theory with stochastic analysis to derive fair prices for options and derivatives. The results are based on stochastic modelling of financial assets, which is the field of *empirical finance*. Numerical approaches for finding prices of options are studied in *computational finance*. All three directions are presented in this book. Algorithms and code for Visual Basic functions are included in the numerical chapter to inspire the reader to test out the theory in practice.

The objective of the book is *not* to give a complete account of option theory, but rather relax the mathematical rigour to focus on the ideas and techniques. Instead of going deep into stochastic analysis, we present the intuition behind basic concepts like the Itô formula and stochastic integration, enabling the reader to use these in the context of option theory. To comprehend the theory, a background in mathematics and statistics at bachelor level (that means, calculus, linear algebra and probability theory) is recommended.

This book is a revision of the Norwegian edition which appeared in 2001. It is used in a course for students at the University of Oslo preparing for a master in finance and insurance mathematics. The manuscript for the Norwegian edition grew from lecture notes prepared for an introductory course in modern finance for the industry.

Several people have contributed in the writing of this book. I am grateful to Jūratė Šaltytė-Benth for carefully reading through the manuscript and significantly improving the presentation, and Neil Shephard for providing me with Ox software to do statistical analysis of financial time series. Furthermore, the advice given and corrections made by Jeffrey Boys, Daniela Brandt,

Catriona Byrne and Susanne Denskus at Springer are acknowledged. All remaining errors are of course the responsibility of the author.

Oslo, August 2003 *Fred Espen Benth*

Table of Contents

1 Introduction 1
  1.1 An Introduction to Options in Finance 1
      1.1.1 Empirical Finance 5
      1.1.2 Stochastic Finance 6
      1.1.3 Computational Finance 6
  1.2 Some Useful Material from Probability Theory 6

2 Statistical Analysis of Data from the Stock Market 11
  2.1 The Black & Scholes Model 12
  2.2 Logarithmic Returns from Stocks 15
  2.3 Scaling Towards Normality 19
  2.4 Heavy-Tailed and Skewed Logreturns 20
  2.5 Logreturns and the Normal Inverse Gaussian Distribution .. 23
  2.6 An Alternative to the Black & Scholes Model 28
  2.7 Logreturns and Autocorrelation 28
  2.8 Conclusions Regarding the Choice of Stock Price Model ... 31

3 An Introduction to Stochastic Analysis 33
  3.1 The Itô Integral 33
  3.2 The Itô Formula 38
  3.3 Geometric Brownian Motion as the Solution of a Stochastic Differential Equation 44
  3.4 Conditional Expectation and Martingales 46

4 Pricing and Hedging of Contingent Claims 53
  4.1 Motivation from One-Period Markets 54
  4.2 The Black & Scholes Market and Arbitrage 58
  4.3 Pricing and Hedging of Contingent Claims $X = f(S(T))$ ... 60
      4.3.1 Derivation of the Black & Scholes Partial Differential Equation 60
      4.3.2 Solution of the Black & Scholes Partial Differential Equation 63
      4.3.3 The Black & Scholes Formula for Call Options 65
      4.3.4 Hedging of Call Options 67
      4.3.5 Hedging of General Options 70

            4.3.6 Implied Volatility................................. 72
    4.4 The Girsanov Theorem and Equivalent Martingale Measures . 73
    4.5 Pricing and Hedging of General Contingent Claims 77
            4.5.1 An Example: a Chooser Option 79
    4.6 The Markov Property and Pricing of General Contingent
         Claims 81
    4.7 Contingent Claims on Many Underlying Stocks............. 83
    4.8 Completeness, Arbitrage and Equivalent Martingale Measures 86
    4.9 Extensions to Incomplete Markets 88
            4.9.1 Energy Markets and Incompleteness 91

5 **Numerical Pricing and Hedging of Contingent Claims** 99
    5.1 Pricing and Hedging with Monte Carlo Methods............ 99
            5.1.1 Pricing and Hedging of Contingent Claims with
                   Payoff of the Form $f(S_T)$........................ 100
            5.1.2 The Accuracy of Monte Carlo Methods 104
            5.1.3 Pricing of Contingent Claims on Many Underlying
                   Stocks 105
            5.1.4 Pricing of Path-Dependent Claims.................. 107
    5.2 Pricing and Hedging with the Finite Difference Method 112

A **Solutions to Selected Exercises** 121

References.. 157

Index.. 161

1 Introduction

1.1 An Introduction to Options in Finance

Suppose you are the risk manager of a pension fund invested in the financial market and you know that in T years the fund needs to pay out €K million in retirement money to the investors. If your fund consists of many risky investments like stocks, the value could be more than €K million, but also less. A possible way to protect the fund is to enter into a contract that guarantees a minimal value of €K million for your assets in T years. Such a contract gives you the right to sell the pension fund at a guaranteed price of €K million in T years time, but if the fund is worth more you are not obliged to do so. Your counterpart, however, is committed to buying your portfolio if its market value is less than €K million. You have entered into a financial contract called a (European) *put option*.

If your portfolio has a market value of less than €K million in T years, you exercise your option. If, on the other hand, the market trades your portfolio for a value higher than €K million in T years, you do not use your right because this would mean selling your pension fund cheaper than necessary. The put option contract gives you the payoff $K - S(T)$ at time T if the market value of the portfolio $S(T)$ is less than K, and zero otherwise. The agreed price K is known as the *strike price*, while T is the *exercise time* of the option contract. Mathematically we write the payoff at the exercise time as

$$\max\left(0, K - S(T)\right). \qquad (1.1)$$

The function $\max(0, x)$ is equal to the maximum of x and zero. Your counterpart in this contract will, however, not enter such a put option without receiving a premium from you. In order to guarantee the value of your pension fund, you have to pay your business partner a certain premium at the time when you buy the put option from her. What should this premium be in order for both of you to accept this deal as fair?

There exist many other types of financial deals with built-in optionalities. The classical example is a (European) *call* option: a producer of steel, say, is dependent on buying energy, and needs protection against too-high energy prices in order to produce steel with profit. If the producer buys a contract that gives the right, but not the obligation, to buy a certain amount of energy for €K at a future date, in T years say, he can control his risk towards high prices in the supply of energy. If the market price of energy $S(T)$ in T years is higher than €K, the producer exercises his option contract and receives energy for a price lower than what is offered in the market. In the opposite case the producer throws away the option contract and goes to the market to buy energy. The payoff from the call option is $S(T) - K$ in the former case, and zero in the latter, or, mathematically,

$$\max\left(0, S(T) - K\right). \tag{1.2}$$

This contract has a strike price K at the exercise time T. What must the steel producer pay to enter into such a call option?

An option contract is a financial asset where the price is dependent on another financial asset. The name *option* indicates that the contract has specific choices or alternatives built in. Such contracts are also called *derivatives*, because their value is *derived* from an *underlying* asset. A huge number of derivatives are traded in today's financial markets, where the underlying asset can be stocks, commodities, bonds, exchange rates, energies etc. We have already mentioned two types of options, namely European call and put options, but optionality can be of many different types. American options give the owner the right to choose the time for exercise; Asian options are options on the average of the underlying asset over a specified time period, and can again be of American or European type. One trades in barrier options, which become worthless if the value of the underlying asset breaks a specified barrier. The introduction of new options and derivatives in the market seems to be endless, and only the imagination of the market participants places bounds on this development. The reader is referred to Hull [30] for a comprehensive presentation of today's derivatives markets.

The main problem we are facing is: what is the price of a derivatives contract? We can rephrase this problem as: what should the buyer pay the seller for such a contract, or what premium is the seller willing to accept in order to commit to a derivatives contract? Mathematical finance gives a solution to this problem and in this book we will try to introduce the reader to the essential parts of the theory behind it. We will develop the theory which provides us with a price for derivatives which both parties in the transaction find acceptable. Moreover, we will also show how the seller of the derivative can hedge away the risk in her short position.

If we have bought a put option on a stock,[1] we do not know what this contract will pay at the time of exercise. The reason is of course that we do not know with certainty the price of the stock at the time of exercise. In Fig. 1.1 the profit from a put option as a function of the stock price at the time of exercise is displayed. The small initial premium paid at the entry of

---

[1] Even though we will mostly use stocks as the underlying asset from now on, the theory presented in this book is also valid for other types of assets.

the put option is included. Figure 1.2 shows the similar profit function for a call option on a stock.

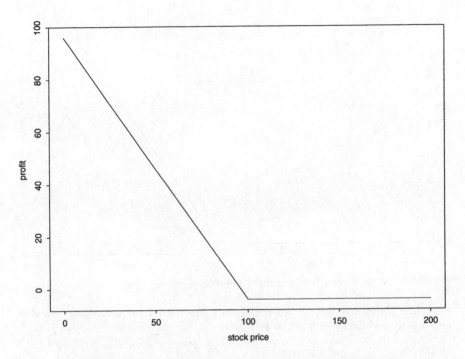

Fig. 1.1. The profit from a put option as a function of the underlying stock at exercise. The strike price is set to $K = 100$

How much a call or put option will be worth to the owner depends on the price of the underlying stock at the exercise time. As we already know, this price is impossible to predict with certainty, but what we can do is to indicate *probable* values of the price of the underlying stock. For instance, based on historical observations of the price fluctuations of a stock, we can build a probabilistic model for the price dynamics and do statistical inference. This is one of the main topics in *empirical finance*. If we have a probabilistic description of the price of the underlying stock at the option's exercise time, a natural guess for the option premium could be the discounted expected payoff from the contract. Such a price is derived from the present-value idea in economics, and for a call option it reads

$$P(0) = (1 + r/100)^{-T} \mathbb{E}\left[\max(0, S(T) - K)\right], \qquad (1.3)$$

where r is the yearly interest rate quoted in percent and $\mathbb{E}$ denotes the expectation. We use the notation $P(0)$ to indicate that this is the price when

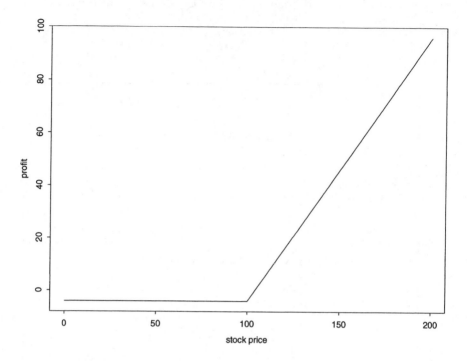

Fig. 1.2. The profit from a call option as a function of the underlying stock at exercise. The strike price is set to $K = 100$

entering into the contract. The expectation can now be calculated appealing to the probabilistic model of $S(T)$. Unfortunately, the price $P(0)$ will in general allow for *arbitrage*: if someone is willing to trade the option contract for this price, it is possible to speculate in the underlying stock and the option contract that cashes-in a certain profit from a zero investment. In this way one can create a money machine, which for obvious reasons cannot exist in any normal market.[2]

With the aid of the mathematical discipline of *stochastic analysis* we are going to derive the price dynamics for options and derivatives that does not leave room for any arbitrage opportunities in the market. Furthermore, we will find the so-called *hedging strategy*, which makes the seller immune to the risk carried by the option. Analysis of pricing and hedging is at the heart of *stochastic finance*.

---

[2] Probably you would detect if your colleague found such a money machine, and start to compete with her. But someone would detect you, and so on. Finally you would all reach an equilibrium price, which, as we will show, is exactly the price that excludes arbitrage possibilities.

1.1 An Introduction to Options in Finance

It turns out that the price of a call option, say, can be written as (1.3), however, with the expectation taken under a different probability than the one used to build the price model for $S(T)$. One needs to work out the option price in a "parallel universe", where the underlying stock is modelled with respect to the *risk-neutral probability*. In this "universe", the expected return from the underlying stock will equal the risk-free interest rate, which means no risk premium. In other words, to find the option premium we need to operate in a probability universe for which the stock price has slightly different properties than in the real world. In many cases one can calculate the expectation and reach analytical formulas, like for call and put options where the price can be derived from the famous Black & Scholes formula. However, most options can only be evaluated by numerical techniques implemented on a computer. The development of such techniques is the topic of *computational finance*. In this book we will pass through all the three fields of mathematical finance; *empirical, stochastic* and *computational* finance.

1.1.1 Empirical Finance

In order to derive the price of an option, we need a stochastic model for the dynamical behaviour of the underlying stock. In particular we need to understand the probabilistic properties of the stock at the exercise time of the option. We have to invent a stochastic model that reflects the observations of the stock's past history as well as possible from a statistical point of view. At the same time, the stochastic model has to fit into a mathematical framework that enables us to analyse option prices. We search for models that both capture the statistical properties of the stock price dynamics and fit into the theory of stochastic analysis. We shall see that the Black & Scholes model for stock dynamics, also known as *geometric Brownian motion* or *the lognormal process*, is a compromise between the two requirements which works reasonably well. The stochastic analysis with respect to the Black & Scholes model becomes simple and in many cases permits explicit formulas.

The task of empirical finance is to investigate financial data from a statistical point of view. In Chap. 2 we shall focus our attention on statistical modelling of data observed in the stock markets, and analyse them with a view towards the Black & Scholes model. The statistical possibilities and limitations of the model will be described, and alternatives will be suggested which fit the stylized facts of financial data better. In particular, we will introduce the geometric normal inverse Gaussian Lévy process as a very flexible and natural class in a financial context. This model was suggested and studied in detail by Barndorff-Nielsen [4]. Even more complex dynamics modelling variations in the volatility, say, lie beyond the scope of this book. However, we will touch on some of the issues leading to the demand for stochastic volatility models, and indicate the consequences for derivatives pricing.

1.1.2 Stochastic Finance

To derive the fair price of an option one constructs a portfolio which replicates perfectly the option's payoff. Such a portfolio is called *the hedging portfolio*, and is an investment in a bond and the underlying stock. The price of the replicating portfolio should be the same as the price of the option, otherwise arbitrage can be constructed by entering into an investment in the hedge and the option.

In Chap. 4 we will show how to construct the replicating portfolio for a general class of option contracts and express the price as a solution of a certain partial differential equation. The solution to this equation can be stated in terms of an expectation, which links the arbitrage-free price of the option to the present value under the risk-neutral probability. Furthermore, for an even more general class of options, so-called contingent claims, we will use the *martingale property* of the hedging portfolio to reach the same pricing formula. All our considerations will be founded on the stochastic analysis presented in Chap. 3.

1.1.3 Computational Finance

If it is not possible to derive an explicit pricing formula for an option contract, one must resort to numerical techniques. Unlike call and put options, most other derivatives do not have an analytical pricing formula, and in Chap. 5 we will introduce two different numerical approaches to solve this problem. Since the price can be derived as the solution of a partial differential equation, one can approximate it numerically by the *finite difference method*. This is a well-known and classical technique for the numerical solution of partial differential equations. On the other hand, the price can also be written as the expected value of the option's payoff (where expectation is with respect to the risk-neutral probability), and this suggests a *Monte Carlo* approach. If we can simulate the payoff from the option in the risk-neutral universe, a simple arithmetic average will yield the price of the option. Both methods also provide the hedging portfolio, as we will show. Chapter 5 includes an implementation of several algorithms in Microsoft's Visual Basic language. The reader is encouraged to try out these methods in order to gain more intuition and knowledge on pricing and hedging of options.

1.2 Some Useful Material from Probability Theory

In this final section we introduce some basic notions from probability theory that will be used throughout the book.

Denote by Ω the probability space (or the sample space) and by $\mathcal{P}$ the probability. If $A \subset \Omega$ is an event then $\mathcal{P}(A)$ is a number in the closed interval $[0, 1]$ giving the probability that A occur. In finance, the probability $\mathcal{P}$ is

1.2 Some Useful Material from Probability Theory

frequently referred to as the *objective* probability, since this gives the probabilities of all events observed in the financial market under consideration. In later chapters we will also construct a so-called *risk-neutral probability*, as mentioned above.

A random variable X is a function from Ω into $\mathbb{R}$, the set of real numbers. This means that for each outcome $\omega \in \Omega$, $X(\omega)$ is a real number. The collection of all numbers $X(\omega)$ is the *state space* of X. The cumulative distribution function of X, P_X, is defined as

$$P_X(x) = \mathcal{P}(X \leq x),$$

that is, the probability that X is less than or equal to the real number x. The α-*quantile*, $0 < \alpha < 1$, of a random variable X is defined as the number q_α such that

$$P_X(q_\alpha) = \mathcal{P}(X \leq q_\alpha) = \alpha.$$

A random variable X is called *normal* (or *Gaussian*) with parameters μ and σ^2 if its cumulative distribution function is

$$P_X(x) = \frac{1}{\sqrt{2\pi\sigma^2}} \int_{-\infty}^{x} \exp\left(-(y-\mu)^2/2\sigma^2\right) dy.$$

We denote such a random variable by $X \sim \mathcal{N}(\mu, \sigma^2)$. When $\mu = 0$ and $\sigma^2 = 1$, X is called a *standard* normal variable, and we will denote its cumulative distribution function by $\Phi(x)$. If the random variable Y can be written as the exponential of a normal random variable X, $Y = \exp(X)$, we say that Y is a *lognormal* random variable. The motivation for this name comes of course from the fact that $\ln(Y) = X$.

The probability density (or simply the density) of a random variable X is defined as the derivative of the cumulative distribution function P_X, that is,

$$p_X(x) = P'_X(x) = \frac{d}{dx}\mathcal{P}(X \leq x).$$

The expectation (or mean) of $g(X)$, where g is a function from $\mathbb{R}$ into $\mathbb{R}$, is

$$\mathbb{E}[g(X)] = \int_{-\infty}^{\infty} g(x) p_X(x) \, dx,$$

which makes sense only as long as this integral is finite. The variance is defined by

$$\text{Var}[X] = \mathbb{E}[X^2] - (\mathbb{E}[X])^2, \tag{1.4}$$

which in terms of the density of X becomes

$$\text{Var}[X] = \int_{-\infty}^{\infty} x^2 p_X(x) \, dx - \left(\int_{-\infty}^{\infty} x p_X(x) \, dx\right)^2.$$

The reader should be aware that not all random variables have finite variance. In fact, there exist many random variables where even the expectation is infinite. However, in this book we shall not encounter such random variables, even though their relevance to finance has been pointed out by Mandelbrot [36, 37]. The square-root of the variance of X is called the *standard deviation* of X, and denoted by $\mathrm{std}(X)$.

We can find the expectation of a random variable X conditioned on the event $A \subset \Omega$ as

$$\mathbb{E}\left[X \mid A\right] = \mathbb{E}\left[1_A X\right].$$

The random variable 1_A is one as long as $\omega \in A$, and zero otherwise, and is called an indicator function of A. Hence, taking the *conditional expectation* of X on A means to average X only over those samples which belong to the event A.

Given two random variables X and Y, their *joint probability density* is defined as

$$p_{X,Y}(x,y) = \frac{\partial^2}{\partial x \partial y} \mathcal{P}\left(X \leq x, Y \leq y\right).$$

The random variables X and Y are said to be *independent* if for any two intervals A and B on the real line

$$\mathcal{P}\left(X \in A, Y \in B\right) = \mathcal{P}\left(X \in A\right) \mathcal{P}\left(Y \in B\right).$$

Another, more operational way, to state independence between two random variables is through their probability density functions: if the probability density functions of X and Y are p_X and p_Y, resp., then the joint probability density function $p_{X,Y}$ of (X,Y) is $p_{X,Y}(x,y) = p_X(x)p_Y(y)$ if and only if X and Y are independent. A measure for statistical dependence between two random variables X and Y is the *covariance*,

$$\mathrm{Cov}\left(X,Y\right) = \mathbb{E}\left[XY\right] - \mathbb{E}\left[X\right]\mathbb{E}\left[X\right]. \tag{1.5}$$

Another popular measure is the correlation, which is the normalized covariance:

$$\mathrm{corr}\left(X,Y\right) = \frac{\mathrm{Cov}\left(X,Y\right)}{\mathrm{std}(X)\mathrm{std}(Y)}. \tag{1.6}$$

Note that when X and Y are independent, $\mathbb{E}\left[XY\right] = \mathbb{E}\left[X\right]\mathbb{E}\left[X\right]$ (see Exercise 1.5), and hence $\mathrm{Cov}\left(X,Y\right) = 0$ and therefore also $\mathrm{corr}\left(X,Y\right) = 0$. The opposite holds only for normally distributed random variables: if X and Y are two normally distributed random variables with correlation equal to zero, then X and Y are independent.

If X is a standard normal random variable we denote the probability density by ϕ, and find

$$\phi(x) = \Phi'(x) = \frac{1}{\sqrt{2\pi}} \exp\left(-x^2/2\right).$$

Furthermore, it is easily seen that when $X \sim \mathcal{N}(\mu, \sigma^2)$ the probability density is

$$p_X(x) = \frac{1}{\sqrt{2\pi\sigma^2}} \exp\left(-\frac{(x-\mu)^2}{2\sigma^2}\right).$$

In Exercise 1.1 the reader is asked to prove that the expectation of X is μ, while the variance is σ^2.

If $X \sim \mathcal{N}(\mu, \sigma^2)$ and Y is a standard normal random variable, it holds that

$$X \stackrel{\mathrm{d}}{=} \mu + \sigma Y.$$

The equality is *in distribution*, which means that X and $\mu + \sigma Y$ have the same probability distribution, namely $\mathcal{N}(\mu, \sigma^2)$ (see Exercise 1.3). This factorization of a normal random variable into a linear combination of a constant and a standard normal random variable will be useful in the derivation of the Black & Scholes formula and in connection with numerical Monte Carlo evaluations of options.

A useful result in statistics is the *central limit theorem*, which we now state:

Theorem 1.1. *Let $X_1, X_2, \ldots$, be a sequence of independent and identically distributed random variables, all having expectation μ and variance σ^2. Then, for any numbers a and b we have*

$$\lim_{n \to \infty} \mathcal{P}\left(a < \frac{X_1 + \cdots + X_n - n\mu}{\sqrt{n}\sigma} < b\right) = \frac{1}{\sqrt{2\pi}} \int_a^b \exp\left(-\frac{1}{2}y^2\right) \mathrm{d}y.$$

We will appeal to this theorem in our study of Monte Carlo methods.

Assume we want to fit a statistical model to a set of observations. Let $\mathbf{x} = (x_1, \ldots, x_n)'$ be a vector of n independent observations (or drawings) from a random variable X having a parametric probability distribution with density $p_X(\theta; x)$, where θ is the vector of parameters. We use the notation $\mathbf{x}'$ to denote the transpose of the vector $\mathbf{x}$. We find the $\widehat{\theta}$ which "maximizes the probability" that the data is drawn from X by using maximum likelihood estimation. Define the likelihood function

$$L(\mathbf{x}; \theta) = \prod_{i=1}^{n} p_X(\theta; x_i).$$

The maximum likelihood estimator of θ is defined as

$$\widehat{\theta} := \arg\max_{\theta} L(\mathbf{x}; \theta).$$

If X is a normal random variable, the maximum likelihood estimators of μ and σ^2 are

$$\widehat{\mu} = \frac{1}{n}\sum_{i=1}^{n} x_i, \quad \widehat{\sigma^2} = \frac{1}{n}\sum_{i=1}^{n}(x_i - \widehat{\mu})^2, \tag{1.7}$$

respectively. Sometimes one divides by $n-1$ instead of n in the expression for $\widehat{\sigma^2}$ in order to have an unbiased estimator. It is standard to use these two estimators for the empirical mean and variance for any sample. The covariance $\gamma_{X,Y} := \mathrm{Cov}(X,Y)$ between two random variables X and Y can be estimated using

$$\widehat{\gamma}_{X,Y} = \frac{1}{n}\sum_{i=1}^{n}(x_i y_i - \widehat{\mu}_X \widehat{\mu}_Y),$$

where $\mathbf{x} = (x_1, \ldots, x_n)'$ and $\mathbf{y} = (y_1, \ldots, y_n)'$ are vectors of n independent observations drawn from X and Y resp., and $\widehat{\mu}_X$ and $\widehat{\mu}_Y$ are the respective empirical means of X and Y.

The reader unfamiliar with basic statistics and probability theory is advised to take a look in any introductory book on the topic, for instance [43].

Exercises

1.1 Demonstrate for $X \sim \mathcal{N}(\mu, \sigma^2)$ that we have

$$\mathbb{E}[X] = \mu, \quad \mathrm{Var}[X] = \sigma^2.$$

1.2 Very often the variance of a random variable X is defined as

$$\mathrm{Var}[X] = \mathbb{E}\left[(X - \mathbb{E}[X])^2\right].$$

Show that this definition coincides with the one in (1.4). Could you, based on this, think of a way other than (1.5) to define the covariance between X and Y?

1.3 If $X \sim \mathcal{N}(\mu, \sigma^2)$, show that

$$a + bX \sim \mathcal{N}(a + b\mu, b^2\sigma^2).$$

All parameters are real numbers and $\sigma > 0$.

1.4 Derive the probability density of a lognormal random variable.

1.5 Show that

$$\mathbb{E}[XY] = \mathbb{E}[X]\mathbb{E}[X],$$

when X and Y are independent random variables.

1.6 Show that the maximum likelihood estimators of the mean and variance for a normal distribution are given as in (1.7).

2 Statistical Analysis of Data from the Stock Market

Black & Scholes assumed in their seminal work [9] that the returns from the underlying stock are normally distributed. The main part of this chapter will be devoted to testing this normal hypothesis on the distribution of observed stock returns. Our purpose is to highlight the basic assumptions and emphasize the limitations of their model. We will also go into other aspects of the stochastic dynamics of the underlying stock. On the way we shall introduce Brownian motion and Lévy processes, as well as some powerful statistical distributions to model stock returns. This chapter will demonstrate the typical problems one is dealing with in empirical finance.

The results from our empirical investigations will strongly argue against the Black & Scholes model. However, when moving on to option theory it is highly desirable to keep their hypothesis for the underlying stock. The Black & Scholes model opens for explicit derivations of pricing formulas and hedging strategies for many interesting examples of options. Moving beyond their framework will lead us into difficult financial and mathematical problems which require a mathematical knowledge beyond the scope of this book. The interested reader, though, will be given references to literature where these parts of mathematical finance can be studied in more depth (see Sect. 4.9 for a further discussion).

Throughout this chapter we will test the Black & Scholes model on data from London and New York stock exchanges. We are going to analyse the FTSE and NASDAQ indexes.[1] We collected from the Yahoo finance site daily, weekly and monthly closing values ranging from 1 January 1990 up to 31 December 2002. For the FTSE index we have 3283 daily, 679 weekly and 156 monthly observations, while the corresponding numbers for the NASDAQ index are 3280 daily, 678 weekly and 156 monthly observations.

---

[1] To be precise, we analyse the FTSE100 and NASDAQ100 indexes. Since the values of these indexes are found by an averaging over many stocks traded on the respective stock exchanges, they measure the overall performance of the stocks. In some sense the indexes can be considered as "super stocks".

2.1 The Black & Scholes Model

To evaluate the price of an option, we need a model which describes statistically the value $S(T)$ of the underlying stock at the exercise time T. Furthermore, if our aim is to replicate (hedge) the option, we need to have a dynamical model for the price $S(t)$ of the underlying stock at all times t between when the option contract is entered into and the exercise time T. The dynamical model of Black & Scholes is *geometric Brownian motion*, also called the *lognormal process*. It is defined mathematically in the following way: for $0 \leq t \leq T$, let $S(t)$ be given by

$$S(t) = S(0) \exp\left(\mu t + \sigma B(t)\right), \qquad (2.1)$$

where μ is the *drift* of the stock and σ is the *volatility*. The current stock price is $S(0)$ and $B(t)$ is *Brownian motion*. To understand what is Brownian motion, we first need to become familiar with the concept of a *stochastic process*:

Definition 2.1. *A stochastic process* $\{X(t)\}_{t \in [0,T]}$ *is a family of random variables parametrized by time* t, *that is, for each given* $t \in [0, T]$, $X(t)$ *is a random variable.*

For $\omega \in \Omega$, we denote an outcome of the random variable $X(t)$ by $X(t, \omega)$. To avoid cumbersome notation, we will from now on simply write $X(t)$ both for the stochastic process $\{X(t)\}_{t \in [0,T]}$ and the random variable when it is clear from the context which one we have in mind. A Brownian motion $B(t)$ is defined as follows:

Definition 2.2. *Brownian motion* $B(t)$ *is a stochastic process starting at zero, i.e.* $B(0) = 0$, *and which satisfies the following three properties:*

1. Independent increments: *The random variable* $B(t) - B(s)$ *is independent of the random variable* $B(u) - B(v)$ *whenever* $t > s \geq u > v \geq 0$.
2. Stationary increments: *The distribution of* $B(t) - B(s)$ *for* $t > s \geq 0$ *is only a function of* $t - s$, *and not of* t *and* s *separately.*
3. Normal increments: *The distribution of* $B(t) - B(s)$ *for* $t > s \geq 0$ *is normal with expectation* 0 *and variance* $t - s$.

Later we will consider Brownian motions which start at positions other than zero, but for the time being we let $B(0) = 0$ be part of the definition. Observe that property 3 implies property 2, so that property 2 is superfluous. However, if we consider a stochastic process $L(t)$ which satisfies only the first two properties in the definition above, we call $L(t)$ a *Lévy process*. Hence, we see that Brownian motion is a particular case of a Lévy process. The reason for stating properties 2 and 3 for Brownian motion is to fit into the more general definition of a Lévy process. Later in this chapter we shall encounter types of Lévy processes other than Brownian motion.

From the definition of a stochastic process, we see that $B(t)$ is a random variable for each time t. But on the other hand, for each outcome ω from the sample space Ω, we will have a function of time, $t \mapsto B(t, \omega)$. This function will be referred to as the *path* of Brownian motion. For each outcome ω we will have a realization of a path, which we then call a sample path. In Fig. 2.1 we have plotted three (simulated) sample paths of Brownian motion.[2] The

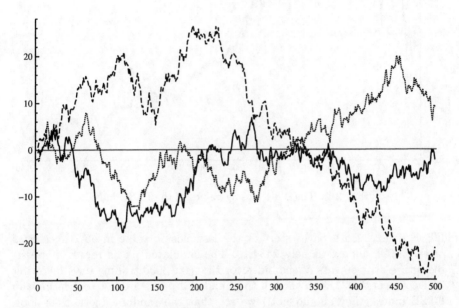

Fig. 2.1. Three simulated paths of Brownian motion

path of a Brownian motion will always be continuous; however, it is a peculiar fact that it is nowhere differentiable. It is worth noting that Brownian motion is the only Lévy process which has continuous paths.

It follows from property 3 of Brownian motion that $B(t)$ is a normal random variable with expectation 0 and variance t (choose $s = 0$). Further, we observe that when $t > s \geq 0$, $B(t) - B(s)$ has the same distribution as $B(t - s)$. Therefore, the expression in the exponent of (2.1), $\mu t + \sigma B(t)$, becomes a normal random variable with expectation μt and variance $\sigma^2 t$, which implies that $S(t)$ is a lognormal random variable – hence the name *lognormal process*. In Fig. 2.2 we have plotted the daily closing values of the FTSE index. Note how the index (and thus the London stock market as a whole) has fallen dramatically in the last few years up to 2003. Figure 2.3 exhibits two simulated paths of a geometric Brownian motion which has been fitted to the

---

[2] See Exercise 5.5, where the task is to develop an algorithm for simulating a path of Brownian motion.

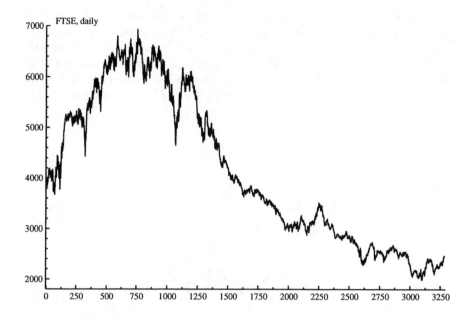

Fig. 2.2. The daily closing values of the FTSE index

FTSE index.[3] Both paths start at the last closing value in our time series of the FTSE index, namely 2434.10. The simulated paths represent possible future evolvements of the index for the next 3000 trading days following 31 December 2002. Even though the simulated paths are not replicating the FTSE index (they should not!) we see that the random fluctuations look somewhat similar.[4] Worth noting is that one of the simulated paths continues to go down, much in accordance with how the market decreased up to the year 2003.

If we let $\sigma = 0$, the stock price dynamics becomes equal to $S(t) = S(0) \exp(\mu t)$. This suggests that the stock grows like the value of a bond yielding a continuously compounding[5] rate of return μ. In common language, one can say that the Black & Scholes model assumes the stock price to vary

---

[3] We will learn how to do this fitting in Sect. 2.2.

[4] The FTSE index seems to be much smoother in its downward phase than the simulated paths, but this is more of an optical illusion since the vertical axis does not extend over the same interval of values.

[5] The actuarial way to calculate interest rates is more common than the continuously compounding: assume a bond pays a yearly interest r_a, then an investment of €1 in the bond will pay back $(1 + r_a)^T$ after T years. The continuously compounded interest rate is then calculated as the number r which solves $\exp(rT) = (1+r_a)^T$, or equivalently, $r = \ln(1+r_a)$. It is a notational advantage to use continuously compounding interest rates when modelling the stock prices as geometric Brownian motions.

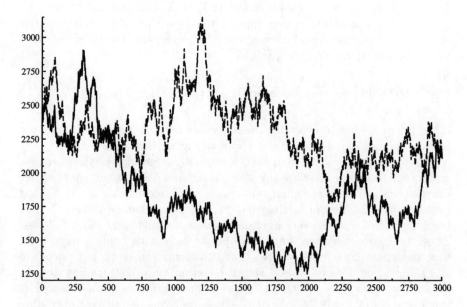

Fig. 2.3. Two simulated paths of a geometric Brownian motion fitted to the FTSE index. Parameters are $\mu = -0.00015$ and $\sigma = 0.01059$

dynamically like the price of a bond yielding a continuously compounding rate of return μ that is randomly distorted by $\sigma B(t)$.

In the rest of this chapter we are going to do an empirical study of stock prices observed in the market and compare with the predictions of the Black & Scholes model. We will try to emphasize the shortcomings of the Black & Scholes model and introduce some new models which are statistically superior. Once again we underline, however, that we are going to use the Black & Scholes model in the subsequent chapters due to its mathematical tractability within the option context. Geometric Brownian motion is the basic model assumption of Black & Scholes, and despite its obvious drawbacks it plays an important role in the modelling of stock prices.

2.2 Logarithmic Returns from Stocks

Assume we have a series of observed stock prices quoted at the times $t_0, t_1, t_2, \ldots, t_N$. Denote the observed stock price at time t_i as $s(i)$. The return at time t_i from an investment in the stock at time t_{i-1} is given by

$$y(i) = \frac{s(i) - s(i-1)}{s(i-1)}, \quad i = 1, \ldots, N.$$

In light of the Black & Scholes model (2.1) it is more natural to consider the so-called *logarithmic return*, that is, the logarithm of the relative price change over the time period in question. The logarithmic return at time t_i of an investment at time t_{i-1} is given by

$$x(i) = \ln\left(\frac{s(i)}{s(i-1)}\right) = \ln(s(i)) - \ln(s(i-1)), \quad i = 1, \ldots, N.$$

If the prices are not too volatile, the difference between returns $y(i)$ and the corresponding logarithmic returns $x(i)$ is negligible (see Exercise 2.5).

The logarithmic return from a stock is usually referred to as the *logreturn*, and we will from here on adopt this shortened name. We assume that the price observations are made at equidistant times, which means that $t_i - t_{i-1} = \Delta t$ for a constant Δt. Measuring time in days, say, means that we observe the last trading price of a stock every day, and remove weekends and other holidays. Thus, we ignore calendar time, and operate only with trading days, with time increment $\Delta t = 1$. Very often time is measured in years, but prices are still sampled on a daily basis. Using the convention of 252 trading days in a year,[6] we have $\Delta t = 1/252$, and thus the *daily* prices $s(i)$ are observed at times $t_i = i/252$, $i = 0, 1, 2, \ldots$. At $t_{252} = 1$, we have one full year of trading days. Contrary to returns $y(i)$, the logreturns $x(i)$ are *additive* in the sense that the sum of n subsequent logreturns is equal to the logreturn over the whole time period, that is,

$$x(i) + \ldots + x(i+n-1) = \ln\left(\frac{s(i+n-1)}{s(i-1)}\right)$$
$$= \ln(s(i+n-1)) - \ln(s(i-1)).$$

We now turn to the question of the statistical properties of the logreturns described by the geometric Brownian motion. To answer this, let us first do a logarithmic transformation of $S(t)$ in (2.1) and look at the increment over one time period

$$X(t_i) := \ln\left(\frac{S(t_i)}{S(t_{i-1})}\right) = \mu \Delta t + \sigma(B(t_i) - B(t_{i-1})), \quad i = 1, 2, \ldots.$$

Recalling the definition of Brownian motion, we have that the increments $B(t_i) - B(t_{i-1})$ for $i = 1, 2, \ldots$ are independent and normally distributed random variables with zero expectation and variance equal to $t_i - t_{i-1} = \Delta t$. Multiplying each increment by a constant σ and adding $\mu \Delta t$ implies that $X(t_i)$ is normally distributed with expectation $\mu \Delta t$ and variance $\sigma^2 \Delta t$. Furthermore, $X(t_i)$ is independent of $X(t_j)$ as long as $i \neq j$. Since the $X(t_i)$'s are independent and identically distributed (or *iid*, as one says in probability

---

[6] This is of course only an average number used for convenience. In practice, each year differs by plus/minus a few days.

theory), the statistical link to the observed logreturns $x(i)$ is particularly simple. In fact, under the Black & Scholes hypothesis, the $x(i)$'s are merely independent drawings from a normal random variable with parameters $\mu \Delta t$ and $\sigma^2 \Delta t$. Hence, it becomes a straightforward task to estimate μ and σ using the maximum likelihood technique. Observe that no matter which time scale we choose, days, weeks or months, the logreturns are always normally distributed under the Black & Scholes paradigm.

Having N logreturn data $x(1), x(2), \ldots, x(N)$, we estimate the expectation μ and variance σ^2 using

$$\widehat{\mu} = \frac{1}{N\Delta t} \sum_{i=1}^{N} x(i), \quad \widehat{\sigma^2} = \frac{1}{(N-1)\Delta t} \sum_{i=1}^{N} (x(i) - \widehat{\mu})^2.$$

Figure 2.4 exhibits the daily logreturns for the FTSE and NASDAQ stock price indexes. A Gaussian kernel smoother is used as a non-parametric estimator of the empirical densities. In the same plot we have included the normal distributions with parameters estimated using the maximum likelihood approach. We see that for both indexes the normal distribution is too wide in the centre compared with the empirical distribution. This is reflected in a lower tail probability than observed from the data. To emphasize the difference, we have plotted the same distributions with a logarithmic scale on the vertical axis in Fig. 2.5. The density function of a normal distribution becomes the parabola

$$-\frac{1}{2}\ln(2\pi\sigma^2) + \frac{(x-\mu)^2}{\sigma^2},$$

when plotted with a logarithmic vertical axis. We see from Fig. 2.5 that the tails of the empirical distributions are not very well fitted by the normal distribution. The estimated parameters are found in Table 2.1.

Table 2.1. Estimated parameters of the normal distribution for daily FTSE and NASDAQ indexes

	$\widehat{\mu}$	$\widehat{\sigma^2}$
FTSE	−0.00015	0.01059^2
NASDAQ	−0.00045	0.02061^2

If we work with time measured in days, we have $\Delta t = 1$ as long as we use prices $s(i)$ observed daily. If, however, time is measured in years, $\Delta t = 1/252$ for daily observations, while $\Delta t = 1/52$ for weekly.[7] The scaling of Δt alters significantly the magnitudes of μ and σ^2. For instance, measuring

---

[7] Assuming 52 trading weeks in a year.

18   2 Statistical Analysis of Data from the Stock Market

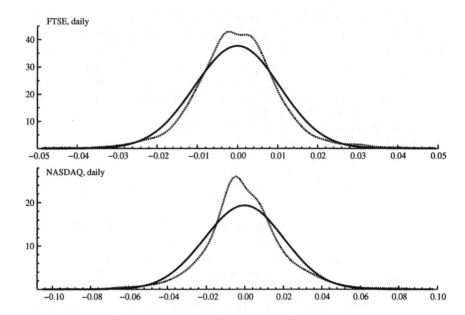

Fig. 2.4. The empirical density (*dotted lines*) and the fitted normal distribution (*solid lines*) of the logreturns of FTSE and NASDAQ indexes

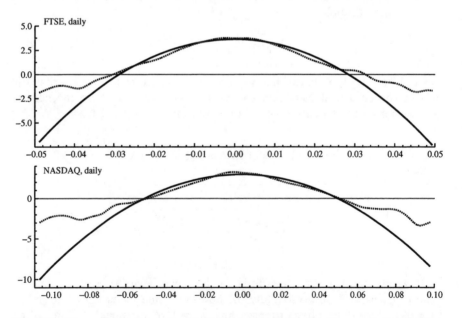

Fig. 2.5. The empirical density (*dotted lines*) and the fitted normal distribution (*solid lines*) of the logreturns of FTSE and NASDAQ indexes with a logarithmic scale on the vertical axis

time in days and observing prices daily yields the estimate $\widehat{\mu} = -0.00045$ and $\widehat{\sigma^2} = 0.02061^2$ for the NASDAQ index. Multiplying these numbers by 252, the corresponding etimates for daily observed data are $\widehat{\mu} = -0.1134$ and $\widehat{\sigma^2} = 0.3271^2$ when time is measured in years. It is fairly common to present the estimated μ and σ in precent using year as the time unit. So, our daily data gives a yearly estimated expected logreturn of about -11.3% with volatility 32.7%.

We end this section by noting that Excel has two built-in functions that do the maximum likelihood estimation for us directly. The function average() finds the average of numbers specified in a row or column, while var() calculates the variance. We can therefore use average() to calculate $\widehat{\mu}$ and var() for $\widehat{\sigma^2}$. The reader should remember, however, that it is necessary to divide by Δt whenever the time increment is different from 1. The function stdev() can be used for finding the standard deviation.

2.3 Scaling Towards Normality

Let us investigate how the distribution of the logreturns is changing when stock prices are sampled over different time spans, for instance, on a daily, weekly or monthly basis. From the additivity property of the logreturns discussed in the section above, the weekly and monthly logreturns can be derived from the daily logreturns. Using five trading days in a week, the logreturn $x^w(i)$ for week i is

$$x^w(i) = \sum_{n=1}^{5} x(5(i-1) + n). \qquad (2.2)$$

The index i is measured in weeks, so for example $x^w(2)$ is the logarithmic change in price from end of week 1 until end of week 2. Similarly, we can define the logreturn $x^m(i)$ for month i as

$$x^m(i) = \sum_{n=1}^{20} x(20(i-1) + n), \qquad (2.3)$$

where we have assumed 20 trading days in a month. For example, $x^m(4)$ is the logreturn at the first day of trading in April from an investment in the stock at the first trading day in March.

If we suppose that the daily logreturns are independent samples from the same distribution (not necessarily a normal distribution), the central limit theorem[8] will imply that the weekly and monthly logreturns tend to a normal distribution. To what extent the daily logreturns are independent samples

---

[8] See for instance [43] or your favourite introductory book on statistics and probability theory.

from the same distribution is not clear from empirical investigations,[9] but it is a statistical fact that the logreturns get closer to a normal distribution when the time intervals become larger (like months, say). However, it is also an empirical fact that on short time intervals (like days or shorter), the logreturns are far from normal. This will be discussed in the next section.

Figure 2.6 exhibits the empirical density for the FTSE index observed weekly and monthly together with the fitted normal distribution. The parameters of the normal distribution were estimated to be $\widehat{\mu} = -0.00068, \widehat{\sigma^2} = 0.02198^2$ for the weekly observations, and $\widehat{\mu} = -0.00328, \widehat{\sigma^2} = 0.04350^2$ for the monthly.[10] Comparing with Fig. 2.4, one observes a convergence towards a normal distribution. However, the monthly data exhibits a distinct skewness which cannot be described by the normal distribution. In Fig. 2.7 the same densities are displayed using a logarithmic scale on the vertical axis. The skewness in the monthly logreturns becomes more apparent, and we also see that the tails are still heavier than the normal. However, the convergence towards normality is clear. The phenomenon of convergence to a normal distribution on longer time scales is often referred to as *Gaussian aggregation*.

2.4 Heavy-Tailed and Skewed Logreturns

Daily logreturns depart from a normal distribution because of heavy tails and skewness. Figure 2.4 shows this for the FTSE index. In this section we will use quantiles to measure how heavy the tails are for the logreturns of the two indexes we consider. A heavy-tailed distribution will have larger quantiles than the normal because it assigns more probability to extreme events.

Tables 2.2 and 2.3 exhibit the empirical[11] quantiles for the daily logreturns of the FTSE and NASDAQ indexes. The tables include the corresponding quantiles for the normal distribution fitted to these data series. The empirical 2% and 98% quantiles are symmetric for both indexes, while the asymmetry enters for the more extreme 1% and 99% quantiles. However, what is obvious is the huge mismatch between the estimated and empirical quantiles.

For the FTSE index the difference is most apparent for the 99% quantile, while the NASDAQ has much heavier tails than the normal for both 1% and 99% quantiles. There are signs of heavy tails also in the 2% and 98% quantiles for both indexes. The normal distribution consistently underestimates the probability for extreme events.

---

[9] The question of independence will be considered in Sect. 2.7.
[10] The reader should note that we use the *actual* weekly and monthly observations here, and not the ones based on the summation formulas (2.2) and (2.3). The difference is negligible.
[11] The empirical $p\%$ quantile q is such that $p\%$ of the data is smaller than q.

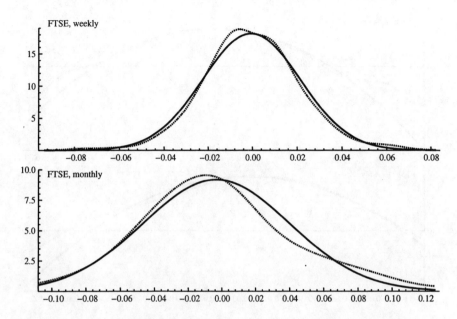

Fig. 2.6. The empirical density (*dotted lines*) of the FTSE weekly (**above**) and monthly (**below**) logreturns plotted together with the fitted normal distribution (*solid lines*)

Table 2.2. Empirical and estimated (normal) quantiles for the daily FTSE index

FTSE	1%	2%	98%	99%
empirical	−0.028	−0.024	0.024	0.030
normal	−0.025	−0.022	0.022	0.025

Table 2.3. Empirical and estimated (normal) quantiles for the daily NASDAQ index

NASDAQ	1%	2%	98%	99%
empirical	−0.060	−0.045	0.045	0.055
normal	−0.048	−0.042	0.042	0.048

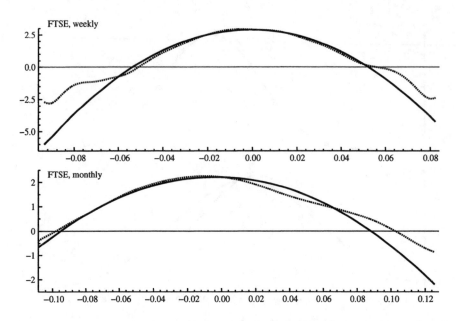

Fig. 2.7. The empirical density (*dotted lines*) of the FTSE weekly (**above**) and monthly (**below**) logreturns plotted together with the fitted normal distribution (*solid lines*) using a logarithmic scale on the vertical axis

How critical is this underestimation? Let us say we would like to find out how many times a year the logreturn of NASDAQ drops more than −0.060. Empirically this happens in 1% of all trading days a year, that is, on average five times in two years.[12] If we believe in the fitted normal distribution, the estimated probability of a logreturn being less than −0.060 is approximately 0.2%. Hence, the fitted normal distribution predicts such extreme events to happen on average only once in two years! If we do the same exercise for the FTSE index, the conclusions are not so dramatic. The fitted normal distribution predicts the probability for a drop in the logreturn bigger than −0.028 to be 0.43%, which roughly means that this happens once a year on average (see Exercise 2.3 for further study).

To get a feeling for how much the skewness has to say, note that the probability assigned by the normal distribution for an increase in logreturn more than the empirical 99% quantile of NASDAQ, is 0.4%, that is, twice as big as the the predicted probability of the empirical 1% quantile. Hence, there is a significant asymmetry between the tails which the normal distribution fails to model.

---

[12] Recall that we assumed 252 trading days in a year.

The notion of *value-at-risk* (or VaR for short) is closely connected with tail probabilities of extreme events in the logreturns. In Exercise 2.4 we define VaR for a stock.

2.5 Logreturns and the Normal Inverse Gaussian Distribution

The normal distribution is not a good model for logreturns on short time intervals. It is not able to model the observed heavy tails, and has no flexibility towards asymmetry and Gaussian aggregation. We are now going to introduce a class of probability distributions which fit the stylized facts of financial time series successfully.

In the literature there exists a huge number of empirical studies on financial logreturn data, and many authors have suggested alternatives to the normal distribution to model these data in a statistically sound way. One of the first to suggest an alternative to the normal paradigm was Mandelbrot in [36] (see also the book [37]). He used the class of stable Pareto distributions to model the returns of cotton and wool prices in the US. The distributions in this class are heavy tailed, but may not be suitable for stock price logreturns since they have infinite variances, a property that seems unrealistic. An alternative class of models that are flexible in modelling financial logreturns is the class of *generalized hyperbolic* distributions. A number of empirical studies have exhibited their ability to model the most important statistical properties observed in financial logreturn data. We refer the reader to the initial works [22] and [4], and for more empirical and theoretical studies, to [45, 48]. In this monograph we are going to present in more detail the normal inverse Gaussian distribution (from here on referred to as NIG), a subclass of the generalized hyperbolic family.

The NIG distribution was introduced by Barndorff-Nielsen [3] in 1977 as a model for the distribution of grain size in sand samples from Danish beaches. Later it found its way into finance and is by now a well-established statistical model for logreturns. The name *normal inverse Gaussian* reflects the fact that the distribution can be represented as a normal mean–variance mixture, that is, a normal distribution with stochastic mean and variance. The mean and variance are both inverse Gaussian distributed. We will not spend time on this representation here, but direct interested readers to [4, 48].

The NIG distribution has four parameters, α, β, μ and δ. The parameter μ describes where the distribution is centred on the real line, while the skewness of the NIG distribution is controlled by β. When $\beta > 0$ the distribution is skewed to the right, while negative β gives skewness to the left. The case of $\beta = 0$ corresponds to a symmetric NIG distribution. The scale parameter is δ, which plays almost the same role as the standard deviation for the normal distribution. The tail heaviness of the distribution is modelled through α.

The probability density of the NIG distribution has the following (slightly cumbersome) expression:

$$p_{\text{nig}}(x;\alpha,\beta,\mu,\delta) = k\exp(\beta(x-\mu))\frac{K_1\left(\alpha\sqrt{\delta^2+(x-\mu)^2}\right)}{\sqrt{\delta^2+(x-\mu)^2}}, \qquad (2.4)$$

where k is the scaling constant $k = \pi^{-1}\delta\alpha\exp(\delta\sqrt{\alpha^2-\beta^2})$ and $K_1(x)$ is the *modified Bessel function of the third kind with index 1*,[13] that is:

$$K_1(x) = \frac{1}{2}\int_0^\infty \exp\left(-\frac{1}{2}x(z+z^{-1})\right)\,dz. \qquad (2.5)$$

The properties of this function are described in detail in [1, Section 9.6]. We see from the definition of the density function in (2.4) that the parameters α and β must satisfy $0 \leq |\beta| \leq \alpha$. In addition, $\delta > 0$.

If the distribution of a random variable L is normal inverse Gaussian with parameters α, β, μ and δ, we write $L \sim \text{NIG}(\alpha,\beta,\mu,\delta)$. The mean and variance of L become

$$\mathbb{E}[L] = \mu + \frac{\delta\beta}{\sqrt{\alpha^2-\beta^2}}, \quad \text{Var}[L] = \frac{\delta\alpha^2}{(\alpha^2-\beta^2)^{3/2}}. \qquad (2.6)$$

Figure 2.8 and 2.9 exhibit four examples of the NIG distribution for different choices of parameters. In Fig. 2.9 all the distributions are plotted using a logarithmic scale on the vertical axis. All the four distributions are centred at the origin with a scale parameter $\delta = 0.015$. The variation of the tail heaviness can be observed in the top rows of the two figures. When $\alpha = 30$ the tails have a hyperbolic shape, while for $\alpha = 150$ they are almost linear, when they are considered on a logarithmic scale. The density of the corresponding normal distribution, that is, the normal distribution with mean and variance equal to that of the NIG in question, will always have parabolically shaped tails when using a logarithmic scale. These figures show the flexibility of the NIG distribution along with the statistical differences from the normal distribution.

Since the density of the NIG distribution is known to us, we can perform maximum likelihood estimation to find the parameters α, β, μ and δ. However, as we can see from the expression for $K_1(x)$ in (2.5), this is a task that does not allow for an explicit solution like the Gaussian case. Bølviken and Benth describe in [8] a numerical procedure for finding the estimators, but the programming language Ox[14] turns estimation into a straightforward exercise since the generalized hyperbolic density is a predefined function. The Ox

---
[13] In the mathematical programming language Matlab they are called the modified Bessel functions of the second kind.
[14] Ox is based on the C/C++ programming language and is documented in [19].

2.5 Logreturns and the Normal Inverse Gaussian Distribution

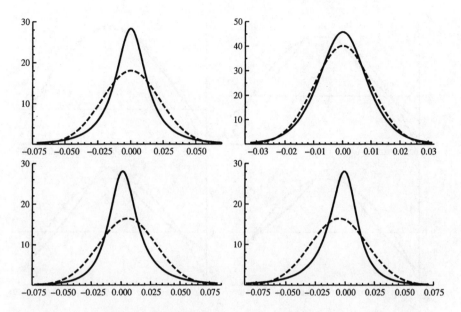

Fig. 2.8. Four different NIG distributions with $\delta = 0.015$ and $\mu = 0$. On the top row the distributions are symmetric with $\alpha = 30$ (**left**) and $\alpha = 150$ (**right**). The bottom row shows two skewed NIG distributions with tail heaviness $\alpha = 30$, and skewness $\beta = 10$ (**left**) and $\beta = -10$ (**right**). The dashed lines show the normal distributions with the same mean and variance as the different NIG distributions

language also supports a tailor-made optimization routine for maximum likelihood estimation. In Tables 2.4 and 2.5 we present the maximum likelihood estimates for the FTSE and NASDAQ indexes. The (rounded) estimates are for daily, weekly and monthly data.

Table 2.4. Maximum likelihood estimation of the NIG distribution for the FTSE index

FTSE	α	β	μ	δ
daily	105	3.0	−0.0005	0.012
weekly	68.2	3.3	−0.0023	0.033
monthly	69.8	26	−0.0465	0.105

Worthwhile mentioning is the strong skewness for monthly data, where the β is 12 for NASDAQ, and 26 for FTSE. The location parameter μ is small as we recognize from the normal distribution. The tail parameter α is in the range 30–40 for NASDAQ, while it is between 68 and 105 for FTSE. In Fig. 2.10 we have plotted the fitted NIG distribution together with the

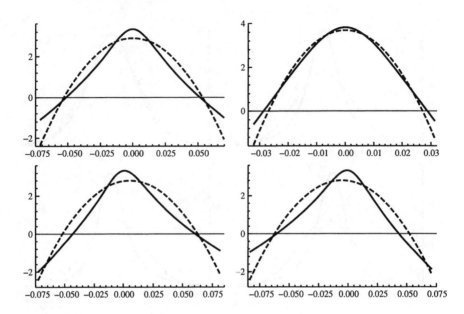

Fig. 2.9. The same as Fig. 2.8 except that logarithmic scale on the vertical axis is used

Table 2.5. Maximum likelihood estimation of the NIG distribution for the NASDAQ index

NASDAQ	α	β	μ	δ
daily	42.3	3.8	−0.0021	0.018
weekly	30.5	5.9	−0.0116	0.048
monthly	31.1	12	−0.0900	0.197

empirical density and the fitted normal density for both daily and monthly data. We notice how the NIG distribution fits the empirical both in the tails and with respect to the skewness, contrary to the normal distribution.

The four parameters of the NIG distribution can be transformed into the so-called *shape triangle* parameters (see [5]):

$$\xi = (1 + \delta\sqrt{\alpha^2 - \beta^2})^{-1/2}, \quad \chi = \xi\beta/\alpha.$$

Since $0 \leq |\chi| < \xi < 1$, the parametrization (χ, ξ) are coordinates in a triangle (inverted). In the limit $(\chi, \xi) \to (0, 0)$, the NIG distribution converges to a Gaussian distribution. When $\chi = 0$, the NIG distribution is symmetric, while $\xi \to 1$ gives the Cauchy distribution in the limit. Table 2.6 gives the (rounded) estimated coordinates for the FTSE and NASDAQ. From these numbers we see that the ξ decreases on longer time scales, a clear sign of the Gaussian

2.5 Logreturns and the Normal Inverse Gaussian Distribution 27

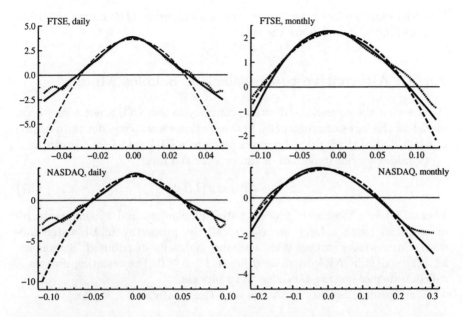

Fig. 2.10. Fitted NIG distribution (*solid line*) together with the empirical density (*dotted line*) and fitted normal distribution (*dashed line*) using a logarithmic scale on the vertical axis

aggregation. Furthermore, only for monthly data we have a χ which is really deviating from zero for FTSE, while it is deviating strongly both for weekly and monthly for NASDAQ. On a daily time scale, the distance to normality is huge. Similar considerations for German, Danish and Norwegian stocks reach the same conclusions (see [48] and [8]).

Table 2.6. Estimated shape triangle parameters for the FTSE and NASDAQ indexes

	FTSE		NASDAQ	
	ξ	χ	ξ	χ
daily	0.67	0.02	0.76	0.07
weekly	0.56	0.03	0.64	0.12
monthly	0.36	0.14	0.39	0.14

We end this section with a result on the convolution of independent NIG distributed random variables. If X and Y are two independent NIG distributed random variables with parameters $(\alpha, \beta, \mu_X, \delta_X)$ and $(\alpha, \beta, \mu_Y, \delta_Y)$ resp., then $X+Y$ is NIG distributed with parameters $(\alpha, \beta, \mu_X+\mu_Y, \delta_X+\delta_Y)$.

The NIG distributions have the convolution property of the normal distributions as long as α and β are the same.

2.6 An Alternative to the Black & Scholes Model

We saw that the normal distribution, contrary to the NIG, is not a very good model for the logreturns of stocks. In this section we will consider the question of what kind of stock price dynamics leads to a NIG model for logreturns.

Consider a price dynamics of the exponential form

$$S(t) = S(0) \exp(L(t)), \qquad (2.7)$$

where $L(t)$ is a stochastic process with independent and stationary increments, thus being a Lévy process (recall the properties of Brownian motion). Furthermore, rather than assuming normally distributed increments, let $L(t) - L(s)$ be NIG distributed for all $t > s \geq 0$. The resulting process is called a *normal inverse Gaussian Lévy* process.

Transforming $S(t)$ to logreturns with time increment 1, we get

$$X(t) = L(t) - L(t-1).$$

The random variables $X(1), X(2), X(3), \ldots$ are independent and NIG distributed since $L(t) - L(t-1)$ has the same distribution as $L(1)$. As in the section above, the parameters α, β, μ and δ of $L(1)$ can be estimated from logreturn data. Using the convolution property of the NIG distributions, we find that $L(t)$ is again a NIG variable with parameters α, β, μt and δt.

In contrast to Brownian motion, the normal inverse Gaussian Lévy process will have discontinuous sample paths. In fact, it is an example of a process with purely discontnuous paths. This implies that the price path of $S(t)$ in (2.7) will not constitute a connected path, but rather jump up and down at arbitrary times. Stochastic and financial analyses for such processes are far more advanced than for Brownian motion (see e.g. [31] and [5]).

2.7 Logreturns and Autocorrelation

The Black & Scholes model (2.1) predicts independent logreturns. If this does not hold true in real markets, one can start to discuss the possibility to predict future returns from historical ones. This sounds very unlikely in liquid markets, and gives a favourable argument for the Black & Scholes model. However, in stock markets one often observes that the price fluctuations cluster into periods with large movements and periods with smaller variations. It therefore seems that the *sizes* of the logreturns may be dependent. This is indeed confirmed empirically, and is what we are going to analyse further in this section.

2.7 Logreturns and Autocorrelation

Consider the time series $\{X(t)\}_{t=1}^{\infty}$ of logreturns and assume it is *stationary* in the sense that the expectation of $X(t)$ is not a function of time and that the covariance between $X(t)$ and $X(t+s)$ only depends on s, the time interval. We write mathematically

$$\text{Cov}(X(t), X(t+s)) = \gamma(s),$$

for a function $\gamma(s)$ where $\gamma(0) := \gamma > 0$. The *autocorrelation* with *lag s* is defined as

$$\text{corr}(X(t), X(t+s)) := \frac{\text{Cov}(X(t), X(t+s))}{\text{std}(X(t))\text{std}(X(t+s))} = \frac{\gamma(s)}{\gamma}.$$

The autocorrelation describes how strongly the current logreturn *remembers* earlier logreturns. If, for instance, $X(t)$ and $X(t+1)$ are positively correlated, $X(t+1)$ will tend to be big every time when $X(t)$ was big. On the other hand, a negative correlation would mean that price drops will more likely be followed by price increases. Thus, if we know the correlation structure in time, we have the tool to predict future prices.

As we have already indicated, it is natural from an economical point of view that the logreturns $X(t)$ are independent. Figure 2.11 validates this empirically for the FTSE index. As we can observe, the logreturns are independent for all lags, since the empirical autocorrelation function is fluctuating randomly around zero. However, if we instead consider the time series $Y(t) = |X(t)|$ or $Z(t) = X^2(t)$, the picture will change. First of all, the absolute value or the square of the logreturns takes away the sign, so we do not consider any longer the size *and* direction of the changes, but only the dependencies between the sizes of the logreturns. In Fig. 2.12 we have plotted the two empirical autocorrelation functions for the FTSE index. The autocorrelation is very small for both $Y(t)$ and $Z(t)$, except at lag 1, where the correlation is a bit below 0.2. However, for all lags we have a consistently positive autocorrelation. This is a sign of what is known as *long-range dependency*, a phenomenon that is frequently observed for logreturns of stocks. Even for big lags there is a positive correlation between the size of the logreturns. Note that *if* the Black & Scholes model were correct, even the absolute values and the squares of the logreturns are independent, and thus have zero autocorrelation for all lags.

In the rest of this book we shall stick to stock price models which predict independent (absolute values and squares of) logreturns, that is, models which produce no autocorrelation for all lags. We see from the discussion above that although such models are empirically wrong, they are not too bad as a first-step approximation. However, if we insist on including time-dependencies in the stock price dynamics, one way out could be to consider stochastic volatility models. Unfortunately, such models lead to complicated (but interesting!) mathematical problems when pricing options, and we will not address these questions here. The interested reader is referred to, e.g., [5].

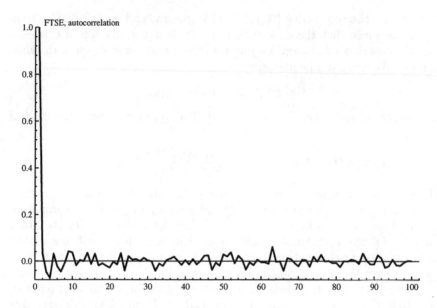

Fig. 2.11. The empirical autocorrelation of the daily FTSE index with lags up to 100

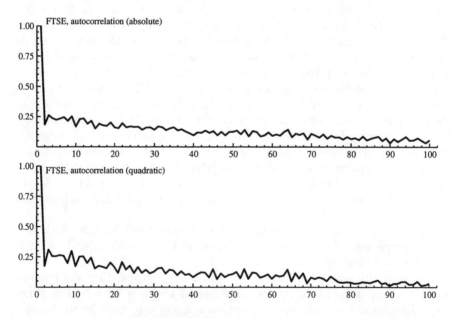

Fig. 2.12. The empirical autocorrelation of the absolute values (**above**) and squares (**below**) of the logreturns from the daily FTSE index with lags up to 100

2.8 Conclusions Regarding the Choice of Stock Price Model

The basic assumption in the Black & Scholes theory for option pricing is the geometric Brownian motion dynamics of the underlying financial stock. This is despite of the empirically observed:

1. heavy tails,
2. skewness, and
3. long-range dependency

for logreturns of financial data. Geometric Brownian motion is not able to model these stylized facts, and other stochastic processes are called for. We have suggested the exponential of a normal inverse Gaussian Lévy process (2.7) as an improvement. The analysis of option premiums based on Lévy processes is much more demanding theoretically than for geometric Brownian motion (see [28],[22] and the discussion in Sect. 4.9).

There are several reasons why one still sticks to the geometric Brownian motion model. First of all, it leads to closed-form solutions of many option premiums. Secondly, the geometric Brownian motion is after all able to model somewhat realistically the dynamical behaviour of many stocks. The shortcomings of the geometric Brownian motion may also become insignificant compared with other market assumptions[15] in the Black & Scholes framework. We shall therefore in the rest of this book assume a geometric Brownian motion dynamics for the price evolution of a stock underlying an option.

Exercises

2.1 Let $B(t)$ be a Brownian motion.
  a) Show that $B(t-s)$ has the same probability distribution as $B(t)-B(s)$ for every $0 \leq s < t$.
  b) Show that $B(t) - B(s)$ is independent of $B(s)$ for every $0 \leq s < t$.

2.2 Calculate the expectation of $S(t)$ given in (2.1). Find also all the moments of $S(t)$.

2.3 Find the probability assigned by the fitted normal distribution for the empirical 2% and 98% quantiles of the NASDAQ index, and the empirical 2%, 98% and 99% quantiles of the FTSE index (all the quantiles can

---

[15] Later, we shall see that Black & Scholes assume for instance that a trader can buy or sell stocks continuously without any transaction costs. Moreover, all options are redundant in the sense that they can be perfectly replicated by speculating in the underlying stock. All these assumptions are critical for option pricing, and may affect the premium much more significantly than the geometric Brownian motion hypothesis.

be found in Tables 2.2 and 2.3). The parameters of the fitted normal distributions are given in Table 2.1.

2.4 Imagine you are a risk manager for a portfolio of stocks, and that you model the value of the portfolio as a geometric Brownian motion $S(t) = S(0)\exp(\mu t + \sigma B(t))$. Your boss has asked you to find the value-at-risk of the portfolio at time t with risk level $\alpha \in (0,1)$. Value-at-risk at time t with risk level α is defined as the number $\text{VaR}_\alpha(t)$ such that

$$\mathcal{P}(S_t \leq \text{VaR}_\alpha(t)) = 1 - \alpha.$$

Hence, there is only a $1 - \alpha$ probability that the portfolio value will be less than $\text{VaR}_\alpha(t)$ at time t. Show that

$$\text{VaR}_\alpha(t) = S_0 \exp\left(\mu t + \sigma\sqrt{t}q_\alpha\right),$$

where q_α is the $1 - \alpha$ quantile of a standard normal distribution.

2.5 Show that $\ln(1+z) \approx z$ when z is small (Hint: use Taylor's Formula with remainder). Use this to show that the logreturn from a stock is approximately equal to the return whenever the latter is small.

2.6 Calculate the kth moment of the increment of a Brownian motion, $B(t+s) - B(t)$, where $s > 0$ and k is a natural number.

3 An Introduction to Stochastic Analysis

Before we can start to explain the option pricing theory of Black & Scholes it is necessary to build up a toolbox of stochastic analysis. This chapter introduces the Itô integral and the Itô formula, which constitute the foundation of stochastic analysis. The martingale property is discussed together with conditional expectation. We restrict our attention to those parts of stochastic analysis that are useful to option theory. Since we try to avoid too advanced mathematics, we shall resort to intuitive arguments rather than being mathematically stringent when presenting the theory. The more theoretically inclined reader is advised to look in [33, 41] for a thorough introduction to stochastic analysis and its many applications.

3.1 The Itô Integral

The Itô integral is at the heart of stochastic analysis, and it is the main reason for the existence of an analysis that differs from the classical mathematical theory of integration and differentiation. The Itô integral defines what one should understand by integration of a stochastic process with respect to Brownian motion (or another stochastic process). The goal of this section is to give an interpretation of the expression

$$\int_0^t X(s)\,dB(s), \tag{3.1}$$

where $X(s)$ is a stochastic process. We say that (3.1) is the Itô integral of $X(t)$ with respect to Brownian motion.

Let us first recall what the interpretation of such an integral would be if $X(s)$ and $B(s)$ *were not* stochastic processes, but rather deterministic functions. Assume $f(s)$ and $g(s)$ are two smooth functions of time s, and consider the integral

$$\int_0^t g(s)\,df(s). \tag{3.2}$$

When $f(s)$ is a differentiable function, we write $df(s)/ds = f'(s)$, or in other words, $df(s) = f'(s)\,ds$. Substituting this into the integral (3.2) leads to

$$\int_0^t g(s)\,\mathrm{d}f(s) = \int_0^t g(s)f'(s)\,\mathrm{d}s,$$

which we recognize as a standard integral.[1] But what happens if $f(s)$ is not differentiable?[2] We can still define the integral (3.2). When $f(s)$ is not fluctuating too much for different arguments s, that is, when $f(s)$ has so-called *bounded variation*, we can prove that the integral is well-defined as the limit

$$\int_0^t g(s)\,\mathrm{d}f(s) = \lim_{n\to\infty} \sum_{i=1}^{n-1} g(s_i)(f(s_{i+1}) - f(s_i)).$$

Since $f(s)$ has bounded variation, $f(s_{i+1})$ is not too far away from $f(s_i)$. From this one can prove that the limit exists as long as $g(s)$ is not varying too much. Of course, if the function $g(s)$ is extremely fluctuating at different points in time, the limit may still diverge.

We shall define the integral (3.1) in an analogous way as the limit

$$\int_0^t X(s,\omega)\,\mathrm{d}B(s,\omega) = \lim_{n\to\infty} \sum_{i=1}^{n-1} X(s_i,\omega)\left(B(s_{i+1},\omega) - B(s_i,\omega)\right). \qquad (3.3)$$

Note that we take the limit for each fixed ω. The problem here is that the ω-wise limit in general does not exist (it becomes $\pm\infty$) for many stochastic processes $X(s)$. For each ω, the function $s \to B(s,\omega)$ is extremely volatile. It is an example of a function which is continuous, but nowhere differentiable. Even worse, Brownian motion as a function of time is not of bounded variation for each ω, as we required for $f(s)$. We need to compensate the roughness of the paths of Brownian motion by putting two conditions on the integrand process $X(s)$. Under these conditions the limit will exist despite the roughness of Brownian motion. The first condition will make $X(s)$ independent of Brownian increments, while the second condition has something to do with the variation of the integrand (similar to the condition that $g(s)$ in (3.2) must not vary too much). Let us go into more detail.

From the third property of Brownian motion we know that the variance of a Brownian increment is given as

$$\mathbb{E}\left[(B(s_{i+1}) - B(s_i))^2\right] = s_{i+1} - s_i.$$

---

[1] Recall that such integrals are defined as a pointwise limit in the following manner:

$$\int_0^t g(s)f'(s)\,\mathrm{d}s = \lim_{n\to\infty} \sum_{i=1}^{n-1} g(s_i)f'(s_i)(s_{i+1} - s_i),$$

where $0 = s_1 < s_2 < \ldots < s_n = t$. Furthermore, we use the convention that the partition points $\{s_1,\ldots,s_n\}$ for n are contained in the partition points $\{s_1,\ldots,s_n,s_{n+1}\}$ for $n+1$. When n increases, we obtain a stepwise refined partition of the interval $[0,t]$.

[2] An example of a function that is not differentiable at point $s = 0$ is $f(s) = |s|$.

If $X(s_i)$ is *independent* of the increment $B(s_{i+1}) - B(s_i)$, we find

$$\mathbb{E}\left[(X(s_i)(B(s_{i+1}) - B(s_i)))^2\right] = \mathbb{E}\left[X^2(s_i)\right] \mathbb{E}\left[(B(s_{i+1}) - B(s_i))^2\right]$$
$$= \mathbb{E}\left[X^2(s_i)\right] (s_{i+1} - s_i).$$

Consider the second moment of the random variable

$$\sum_{i=1}^{n-1} X(s_i) (B(s_{i+1}) - B(s_i)). \tag{3.4}$$

By assuming that $X(s_i)$ is independent of $B(s_{i+1}) - B(s_i)$ for every $i = 1, \ldots, n-1$, we find by independence of the Brownian increments that (see Exercise 3.1):

$$\mathbb{E}\left[\left(\sum_{i=1}^{n-1} X(s_i)(B(s_{i+1}) - B(s_i))\right)^2\right] = \sum_{i=1}^{n-1} \mathbb{E}\left[X^2(s_i)\right](s_{i+1} - s_i).$$

We recognize the sum on the right-hand side as an approximation of the integral $\int_0^t \mathbb{E}\left[X^2(s)\right] ds$. Hence, if this integral exists, we deduce

$$\lim_{n \to \infty} \mathbb{E}\left[\left(\sum_{i=1}^{n-1} X(s_i)(B(s_{i+1}) - B(s_i))\right)^2\right] = \int_0^t \mathbb{E}\left[X^2(s)\right] ds,$$

which yields the conclusion that the variance of the sum in (3.4) converges to $\int_0^t \mathbb{E}\left[X^2(s)\right] ds$. Assuming that this integral exists, we have shown that

$$\mathbb{E}\left[\left(\lim_{n \to \infty} \sum_{i=1}^{n-1} X(s_i)(B(s_{i+1}) - B(s_i))\right)^2\right] = \int_0^t \mathbb{E}\left[X^2(s)\right] ds. \tag{3.5}$$

Let us devote a few lines to discussing the assumption that the right-hand side of (3.5) exists. This integral exists whenever it is finite, that is, whenever the stochastic process $X(s)$ is such that its second moment can be integrated from 0 to t. It is simple to make examples where this is not the case. Take, for instance, the process $X(s) = s^{-1} B(s)$. From Property 3 of Brownian motion, we have

$$\int_0^t \mathbb{E}\left[X^2(s)\right] ds = \int_0^t s^{-1} ds = \ln t - \ln 0 = +\infty.$$

On the other hand, letting $X(s) = B(s)$, the reader can easily check that $X(s)$ satisfies the integrability condition. It holds for a large class of stochastic processes.

Going back to relation (3.5), we see from the limit on the left-hand side that the $X(s_i)$'s have to be independent of the increments $B(s_{i+1}) - B(s_i)$ *for all possible choices of* s_i, $i = 1, \ldots, n-1$. This leads us to the second condition which says that the integrand process has to be an *adapted* process:

Definition 3.1. *A random variable X is called $\mathcal{F}_s$-adapted if X can be written as (a limit of a sequence of) functions of $B(\tau)$ for one or more $\tau \leq s$, but not as a function of any $B(u)$ with $u > s$. A stochastic process $X(s)$ is called adapted if for each time $s \in [0, t]$ the random variable $X(s)$ is $\mathcal{F}_s$-adapted.*

Later the mysterious notation "$\mathcal{F}_s$" will become more sensible. For the time being it will just serve as a cumbersome notation.[3]

Before proceeding, let us elaborate some points of Def. 3.1. First of all, simple processes which are made up of Brownian motion like $X(s) = f(s, B(s))$ are adapted, while a process like $X(s) = B(s+1)$ is not. If we consider the time integral $X(s) = \int_0^s B(\tau) \, d\tau$, it will also define an adapted stochastic process, because the integral is the limit of sums of Brownian motions at different times up to and including s. Namely, by the definition of the integral, we have

$$X(s) = \int_0^s B(\tau) \, d\tau = \lim_{n \to \infty} \sum_{i=1}^{n-1} B(\tau_i) \, (\tau_{i+1} - \tau_i).$$

A similar type of integral appears in connection with so-called Asian options, which are options based on the average of the stock price. In that case we integrate a geometric Brownian motion rather than Brownian motion itself.

Going back to our deduction of the Itô integral, we see that whenever $X(s)$ is an adapted process, the integral $\int_0^t X(s) \, dB(s)$ gives meaning as the pointwise limit in (3.3). As we just showed, the limit in (3.3) converges in variance, and thus also for every $\omega \in \Omega$.[4] We conclude our discussion with the definition of the Itô integral.

Definition 3.2 (Itô integration). *A stochastic process $X(s)$ is called Itô integrable on the interval $[0, t]$ if:*

1. *$X(s)$ is adapted for $s \in [0, t]$, and*
2. *$\int_0^t \mathbb{E}\left[X^2(s)\right] ds < \infty$.*

The Itô integral is defined as the random variable

$$\int_0^t X(s, \omega) \, dB(s, \omega) = \lim_{n \to \infty} \sum_{i=1}^{n-1} X(s_i, \omega) \left(B(s_{i+1}, \omega) - B(s_i, \omega)\right), \quad (3.6)$$

where the limit is taken for every $\omega \in \Omega$.

---

[3] If you know measure theory, $\mathcal{F}_s$ is a σ-algebra, and the notion of adaptedness is the same as measurability with respect to this σ-algebra.

[4] Those who know measure theory will immediately recognize that this is not completely true, but has to be understood as a limit for *almost every* ω. Even more, we only have convergence for a subsequence. The reader who wants to be precise on this point is referred to, e.g. [33].

Notice that the Itô integral itself becomes a stochastic process as it is parametrized by time t. Furthermore, this process is adapted over every time interval since it will be a limit of a sum of functions of Brownian motions $B(s)$ for times $s \leq t$.

In Exercise 3.2 the reader is asked to prove the following theorem.

Theorem 3.3. *The expectation and variance of the Itô integral are*

$$\mathbb{E}\left[\int_0^t X(s)\,\mathrm{d}B(s)\right] = 0, \quad \mathrm{Var}\left(\int_0^t X(s)\,\mathrm{d}B(s)\right) = \int_0^t \mathbb{E}\left[X^2(s)\right]\,\mathrm{d}s. \quad (3.7)$$

The relation for the variance is known as the Itô *isometry.*

We derive some simple properties of the Itô integral which are useful in calculations. First of all, the Itô integral is linear. If $X(s)$ and $Y(s)$ are two Itô integrable processes, then $aX(s) + bY(s)$ is Itô integrable and

$$\int_0^t (aX(s) + bY(s))\,\mathrm{d}B(s) = a\int_0^t X(s)\,\mathrm{d}B(s) + b\int_0^t Y(s)\,\mathrm{d}B(s),$$

where a and b are constants (see Exercise 3.3). Further, the process $X(s) = 1$ is Itô integrable and thus we have from the linearity that for a constant a

$$\int_0^t a\,\mathrm{d}B(s) = a\int_0^t \mathrm{d}B(s) = aB(t).$$

The definition of the Itô integral is not very operational. As the reader recalls from basic calculus, antiderivation is the clue when calculating integrals. In the context of Itô integrals, there exists a similar "antiderivation" technique, namely Itô's formula. This is the topic of the next section where we will apply this formula to demonstrate how to calculate Itô integrals. To see what we can expect when integrating stochastic processes with respect to Brownian motion, we include one example that we will come back to in connection with Itô's formula. Using the limit definition and the property that Brownian motion itself is Itô integrable, it is possible to show that

$$\int_0^t B(s)\,\mathrm{d}B(s) = \frac{1}{2}B^2(t) - \frac{1}{2}t.$$

We obtain an additional term $-t/2$ on the right-hand side, which would not appear if the Itô integral followed classical rules of calculus: if we integrate a differentiable function $f(s)$ with respect to itself, we get (with $f(0) = 0$),

$$\int_0^t f(s)\,\mathrm{d}f(s) = \int_0^t f(s)f'(s)\,\mathrm{d}s = \frac{1}{2}f^2(t).$$

Here we do not get any extra correction term. This integral, which can be easily found after introducing the Itô formula, shows the difference between stochastic analysis and classical rules of calculus. Later we will meet more examples displaying the differences.

3.2 The Itô Formula

Itô's formula is a remedy that can be used to calculate explicitly many Itô integrals. However, the formula has a much wider range of applications, and is one of the main tools to derive prices of option contracts. Itô's formula is a stochastic version of the classical chain rule of differentiation, and prescribes how a function of Brownian motion $f(B(t))$, or more generally, a function of a stochastic process $f(X(t))$, changes stochastically as time progresses. The stochastic dynamics of $f(X(t))$ will be decomposed into the dynamics of the process $X(t)$ and the rate of change of $f(x)$, given by its derivatives. The Itô integral is the main ingredient in the stochastic chain rule. Together with the Itô integral, Itô's formula is the foundation for modern *stochastic analysis*. Let us start the introduction of Itô's formula by recalling the classical chain rule for differentiable functions.

Let $f(t)$ and $g(t)$ be two differentiable functions. By appealing to the chain rule, we find the derivative of $f(g(t))$ with respect to t as

$$\frac{d}{dt}f(g(t)) = f'(g(t))\,g'(t).$$

Integrating both sides with respect to time from 0 to t, and assuming that $g(0) = x$, we find

$$f(g(t)) = f(x) + \int_0^t f'(g(s))\,g'(s)\,ds.$$

This is the integral form of the chain rule. Since each path of Brownian motion is a function of time, one could think that putting $g(t) = B(t)$ would give us the Itô formula as:

$$f(B(t)) = f(x) + \int_0^t f'(B(s))\,B'(s)\,ds.$$

When defining the Itô integral, we made a point of the non-differentiability of the Brownian paths. Hence, $B'(s)$ is not meaningful. However, we could interpret $B'(s)\,ds$ as $dB(s)$, and suggest

$$f(B(t)) = f(x) + \int_0^t f'(B(s))\,dB(s), \tag{3.8}$$

as the Itô formula. Unfortunately, (3.8) is *not* correct.

Before we proceed, note the assumption $g(0) = x$. This means that Brownian motion starts at x rather than zero, $B(0) = x$. As we will see later, it is convenient to have the starting position of the Brownian motion at an arbitrary point x. There are several ways to define a Brownian motion starting at x. Going back to the definition, we could suppose $B(0) = x$, and then in Property 3 let $B(t) - B(s) \sim \mathcal{N}(x, t-s)$. Alternatively, we can define a

3.2 The Itô Formula

Brownian motion starting at x as the process $B^x(t) := x + B(t)$, where $B(t)$ is defined as a Brownian motion starting at zero.

The chain rule is proved by first using a Taylor expansion of $f(g(s_{i+1}))$ around $f(g(s_i))$ where $\{s_i\}_{i=1}^n$ is a partition of the interval $[0,t]$, and then letting the number of partition points go to infinity. Following the same procedure in the case of Brownian motion, one obtains additional correction terms since the variances of Brownian increments are equal to the corresponding time increments. These additional terms do not converge to zero after letting $n \to \infty$, but to some integral over time. We now go into more detail in the derivation of what will become Itô's formula.

Suppose that Brownian motion starts at x, $B(0) = x$, and f is a function which is twice differentiable. Let furthermore s_i and s_{i+1} be two arbitrary points in the interval $[0,t]$ with $s_i < s_{i+1}$. A second-order Taylor expansion around $B(s_i)$ with remainder leads to

$$f(B(s_{i+1})) - f(B(s_i)) = f'(B(s_i))(B(s_{i+1}) - B(s_i))$$
$$+ \frac{1}{2} f''(B(s_i))(B(s_{i+1}) - B(s_i))^2$$
$$+ R\left\{(B(s_{i+1}) - B(s_i))^3\right\}. \tag{3.9}$$

Assume next that we have a sequence of points $\{s_i\}_{i=1}^n$ with the property $0 = s_1 < s_2 < \cdots < s_{n-1} < s_n = t$. A summation from $i = 1$ to $n-1$ gives

$$f(B(t)) - f(x) = \sum_{i=1}^{n-1} f'(B(s_i))(B(s_{i+1}) - B(s_i))$$
$$+ \frac{1}{2} \sum_{i=1}^{n-1} f''(B(s_i))(B(s_{i+1}) - B(s_i))^2$$
$$+ \sum_{i=1}^{n-1} R\left\{(B(s_{i+1}) - B(s_i))^3\right\}.$$

From Def. 3.2 we know that the first term on the right-hand side will converge to the Itô integral when $n \to \infty$. But what about the two other terms? Calculating the expectation of the second term gives

$$\mathbb{E}\left[\sum_{i=1}^{n-1} f''(B(s_i))(B(s_{i+1}) - B(s_i))^2\right] = \sum_{i=1}^{n-1} \mathbb{E}[f''(B(s_i))](s_{i+1} - s_i)$$
$$\longrightarrow \int_0^t \mathbb{E}[f''(B(s_i))]\, ds,$$

when $n \to \infty$. The limit holds as long as the integral is finite, which is assured if, for instance, f'' is a bounded function. Since this integral in general is different from zero, we conclude that the second sum on the right-hand side of (3.9) converges to some non-zero limit. In fact, after some tedious calculations, one can prove that this sum tends to $\int_0^t f''(B(s))\,\mathrm{d}s$. The trick is to show

$$\mathbb{E}\Big[\Big(\sum_{i=1}^{n-1} f''(B(s_i))\big(B(s_{i+1})-B(s_i)\big)^2 - \sum_{i=1}^{n-1} f''(B(s_i))\big(s_{i+1}-s_i\big)\Big)^2\Big] \longrightarrow 0,$$

for $n \to \infty$ (see Exercise 3.4).

Similar considerations reveal that

$$\sum_{i=1}^{n-1} R\Big\{\big(B(s_{i+1})-B(s_i)\big)^3\Big\} \to 0,$$

when $n \to \infty$. The proof of this fact is a bit more messy, and will not be elaborated further (the reader is encouraged to check out [41] for the details). Collecting everything together, we end up with the following result.

Theorem 3.4 (Itô's formula for Brownian motion). *Let $f(x)$ be a twice differentiable function and assume that Brownian motion starts at x. Then,*

$$f(B(t)) = f(x) + \int_0^t f'(B(s))\,\mathrm{d}B(s) + \frac{1}{2}\int_0^t f''(B(s))\,\mathrm{d}s. \tag{3.10}$$

The reader may wonder about additional conditions on f to assure the existence of the integrals on the right-hand side of (3.10). In fact, there is a huge theory on stochastic processes that proves the validity of the formula under the single condition of twice differentiability[5] of f, but then we need to generalize our approach for Itô integration (see, e.g. [33]). We will not follow that path, but instead state some sufficient conditions on f yielding the validity of (3.10). The stochastic process $f'(B(s))$ is Itô integrable whenever the condition $\int_0^t \mathbb{E}\left[f'(B(s))^2\right]\,\mathrm{d}s < \infty$ is satisfied since the process is clearly adapted. The $\mathrm{d}s$ integral in (3.10) is well-defined as long as $f''(B(s))$ is integrable in the classical sense. This is true when, for instance, f'' is a continuous function, because then the stochastic process $f''(B(s))$ will have continuous paths and thereby be a bounded function in the interval $[0,t]$ (see Exercise 3.5). A sufficient condition to have finite variance is $\mathbb{E}\left[\int_0^t f''(B(s))^2\,\mathrm{d}s\right] < \infty$ (see Exercise 3.5). Summing up, we impose two extra conditions on f:

$$\mathbb{E}\left[\int_0^t f'(B(s))^2\,\mathrm{d}s\right] < \infty, \quad \mathbb{E}\left[\int_0^t f''(B(s))^2\,\mathrm{d}s\right] < \infty. \tag{3.11}$$

Note that these two conditions imply that $f(B(t))$ has finite variance. In the future, we shall always assume that these integrability conditions hold. Note

---

[5] To be precise, f must be twice *continuously* differentiable.

that since f is assumed to be twice differentiable, we cannot apply functions like $f(x) = |x|$ in Itô's formula.

As a first application of Itô's formula, let us calculate the integral $\int_0^t B(s)\,\mathrm{d}B(s)$ that we considered in the previous section.

Example 3.5. Let $f(x) = x^2$, and suppose Brownian motion starts at zero. Appealing to the Itô formula for Brownian motion, we find

$$B^2(t) = 0^2 + \int_0^t 2B(s)\,\mathrm{d}B(s) + \frac{1}{2}\int_0^t 2\,\mathrm{d}s,$$

and therefore

$$\int_0^t B(s)\,\mathrm{d}B(s) = \frac{1}{2}B^2(t) - \frac{1}{2}t.$$

For later purposes it is desirable to have the Itô formula for more general stochastic processes than simply Brownian motion. When pricing option contracts, we shall face the problem of finding the dynamics of a function of geometric Brownian motion, and a slight generalization of Itô's formula will simplify this task. The natural class of stochastic processes to consider is the so-called *semimartingales*. A semimartingale is a stochastic process which can be decomposed into an Itô integral and a standard integral. We make the following definition.

Definition 3.6. *The stochastic process $X(t)$ is called a semimartingale if there exist two Itô integrable stochastic processes $Y(t)$ and $Z(t)$ such that*

$$X(t) = x + \int_0^t Y(s)\,\mathrm{d}B(s) + \int_0^t Z(s)\,\mathrm{d}s. \tag{3.12}$$

Note that we assume both processes $Y(t)$ and $Z(t)$ to be adapted, which leads to the adaptedness of $X(t)$. Moreover, since Z also is Itô integrable, the semimartingale has a finite second moment (see Exercise 3.6). If we put $Z(t) = 0$, the semimartingale $X(t)$ reduces to an Itô integral. This special case of a semimartingale is known as a *martingale*, a class of stochastic processes which we will come back to in Sect. 3.4.

In the following theorem we state the general Itô formula.

Theorem 3.7 (General Itô formula). *Assume that $f(t,x)$ is a function which is once differentiable in t and twice differentiable in x, and let $X(t)$ be a semimartingale. Then*

$$f(t, X(t)) = f(0, x) + \int_0^t Y(s)\frac{\partial f(s, X(s))}{\partial x}\,\mathrm{d}B(s) \tag{3.13}$$
$$+ \int_0^t \frac{\partial f(s, X(s))}{\partial t} + Z(s)\frac{\partial f(s, X(s))}{\partial x} + \frac{1}{2}Y^2(s)\frac{\partial^2 f(s, X(s))}{\partial x^2}\,\mathrm{d}s.$$

The derivation of this chain rule for semimartingales follows closely the argument for Brownian motion. We need to impose some additional integrability conditions in order to assure the existence of the different terms. Itô integrability and the existence of the second moments are verified under the conditions (see Exercise 3.7)

$$\mathbb{E}\left[\int_0^t \left(\frac{\partial f(s,X(s))}{\partial t}\right)^2 + Z^2(s)\left(\frac{\partial f(s,X(s))}{\partial x}\right)^2 \right.$$
$$\left. + Y^4(s)\left(\frac{\partial^2 f(s,X(s))}{\partial x^2}\right)^2 ds\right] < \infty, \quad (3.14)$$

and

$$\mathbb{E}\left[\int_0^t Y(s)^2 \left(\frac{\partial f(s,X(s))}{\partial x}\right)^2 ds\right] < \infty, \quad (3.15)$$

which are slightly stronger than (3.11). Note that $f(t, X(t))$ is a semimartingale. In fact, Itô's formula demonstrates that a function of a semimartingale is again a semimartingale.

Normally the Itô formula is written in its *differential form*. To have a more compact notation one suggestively writes

$$df(t, X(t)) = \left\{\frac{\partial f(t,X(t))}{\partial t} + Z(t)\frac{\partial f(t,X(t))}{\partial x} + \frac{1}{2}Y^2(t)\frac{\partial^2 f(t,X(t))}{\partial x^2}\right\} dt$$
$$+ Y(t)\frac{\partial f(t,X(t))}{\partial x} dB(t). \quad (3.16)$$

We shall adopt this way of stating the stochastic chain rule, but warn the reader that this is only meant as a suggestive formulation with the precise meaning stated in Thm. 3.7. It is sometimes useful to use the following shorthand version of (3.16):

$$df(t, X(t)) = \frac{\partial f(t,X(t))}{\partial t} dt + \frac{\partial f(t,X(t))}{\partial x} dX(t)$$
$$+ \frac{1}{2}\frac{\partial^2 f(t,X(t))}{\partial x^2} (dX(t))^2, \quad (3.17)$$

together with the calculation rules $(dt)^2 = 0, dt dB(t) = dB(t)dt = 0$ and $(dB(t))^2 = dt$. Applying these rules for (3.17) we get back (3.16).

Let us pause our theoretical considerations and look at an example.

Example 3.8. Consider the stochastic process $U(t) = e^{-\lambda t}\int_0^t e^{\lambda s} dB(s)$, and suppose that we would like to find its dynamics $dU(t)$. In order to use the Itô formula, we need to identify the semimartingale $X(t)$ along with the function $f(t,x)$. In this case, let $X(t) = \int_0^t e^{\lambda s} dB(s)$ and $f(t,x) = e^{-\lambda t}x$. The process $X(t)$ is obviously a semimartingale starting at zero with $Y(s) =$

$e^{\lambda s}$ and $Z(s) = 0$ (in fact a martingale as we will see later). Hence, since $\partial f(t,x)/\partial t = -\lambda e^{-\lambda t} x$, $\partial f(t,x)/\partial x = e^{-\lambda t}$ and $\partial^2 f(t,x)/\partial x^2 = 0$,

$$dU(t) = -\lambda e^{-\lambda t} X(t)\, dt + e^{-\lambda t}\, dX(t) + \frac{1}{2} \times 0\, (dX(t))^2.$$

Using the fact that $dX(t) = e^{\lambda t}\, dB(t)$ we find

$$dU(t) = -\lambda U(t)\, dt + dB(t).$$

The reader is encouraged to show that the integrability conditions (3.14) and (3.15) are validated for this example (see Exercise 3.8). The process $U(t)$ is known as an Ornstein–Uhlenbeck process and is used in finance as a basic tool for modelling interest rates and commodity prices (see, e.g. [13, 49] for more details).

We finish this section with a multi-dimensional Itô formula. Introduce m independent Brownian motions $B_1(t), \ldots, B_m(t)$, and assume that $X_1(t), \ldots, X_n(t)$ are n semimartingales with dynamics

$$dX_1(t) = Y_{11}(t)\, dB_1(t) + \cdots + Y_{1m}(t)\, dB_m(t) + Z_1(t)\, dt$$
$$\ldots$$
$$\ldots$$
$$dX_n(t) = Y_{n1}(t)\, dB_1(t) + \cdots + Y_{nm}(t)\, dB_m(t) + Z_n(t)\, dt.$$

The notation $\mathbf{X}(t)$ stands for the vector $(X_1(t), \ldots, X_n(t))'$. Consider now a vector-valued function of t and $\mathbf{x} \in \mathbb{R}^n$, $\mathbf{g}(t, \mathbf{x}) = (g_1(t, \mathbf{x}), \ldots, g_p(t, \mathbf{x}))'$. The stochastic dynamics of $\mathbf{g}(t, \mathbf{X}(t))$ is given from the multi-dimensional version of the Itô formula by considering each coordinate process. For $k = 1, \ldots, p$,

$$dg_k(t, \mathbf{X}(t)) = \frac{\partial g_k(t, \mathbf{X}(t))}{\partial t}\, dt + \sum_{i=1}^{n} \frac{\partial g_k(t, \mathbf{X}(t))}{\partial x_i}\, dX_i(t)$$
$$+ \frac{1}{2} \sum_{i,j=1}^{n} \frac{\partial^2 g_k(t, \mathbf{X}(t))}{\partial x_i x_j}\, dX_i(t) dX_j(t), \qquad (3.18)$$

with the rules $dB_i(t)dB_j(t) = \delta_{ij} dt$, $(dt)^2 = dB_i(t)dt = dt dB_i(t) = 0$ and $\delta_{ij} = 1$ when $i = j$, and zero otherwise. Here is an example applying the multi-dimensional Itô formula:

Example 3.9. Assume that $n = m = 2$, and

$$dX_1(t) = \sigma_{11}\, dB_1(t) + \sigma_{12}\, dB_2(t) + \mu_1\, dt$$
$$dX_2(t) = \sigma_{21}\, dB_1(t) + \sigma_{22}\, dB_2(t) + \mu_2\, dt,$$

where μ_i, σ_{ij}, $i,j = 1,2$, are all constants. What is the dynamics of the stochastic process $U(t) = X_1(t)X_2(t)$? Appealing to the multi-dimensional Itô formula with the real-valued function $g(t, x_1, x_2) = x_1 x_2$, gives

$$dg(t, X_1(t), X_2(t)) = X_2(t)\, dX_1(t) + X_1(t)\, dX_2(t) + dX_1(t) dX_2(t).$$

The rules of calculation imply $dX_1 dX_2 = (\sigma_{11}\sigma_{21} + \sigma_{12}\sigma_{22})dt$, and hence

$$dU(t) = (\mu_1 X_2(t) + \mu_2 X_1(t) + (\sigma_{11}\sigma_{21} + \sigma_{12}\sigma_{22}))\, dt$$
$$+ (\sigma_{11} X_2(t) + \sigma_{21} X_1(t))\, dB_1(t) + (\sigma_{12} X_2(t) + \sigma_{22} X_1(t))\, dB_2(t).$$

3.3 Geometric Brownian Motion as the Solution of a Stochastic Differential Equation

Recall geometric Brownian motion as the stochastic process

$$S(t) = S(0) \exp\left(\mu t + \sigma B(t)\right). \tag{3.19}$$

Let us now use Itô's formula to rewrite this process into a semimartingale. Introduce the function

$$f(t, x) = S(0) \exp\left(\mu t + \sigma x\right),$$

and observe that $S(t) = f(t, B(t))$. Since $\partial f(t,x)/\partial t = \mu f(t,x)$, $\partial f(t,x)/\partial x = \sigma f(t,x)$ and $\partial^2 f(t,x)/\partial x^2 = \sigma^2 f(t,x)$, we find from Itô's formula with $X(t) = B(t)$ that

$$df(t, B(t)) = \mu f(t, B(t))\, dt + \sigma f(t, B(t))\, dB(t) + \frac{1}{2}\sigma^2 f(t, B(t))\, (dB(t))^2.$$

Since $(dB(t))^2 = dt$, we are left with the following dynamics of $S(t)$:

$$dS(t) = \left(\mu + \frac{1}{2}\sigma^2\right) S(t)\, dt + \sigma S(t)\, dB(t). \tag{3.20}$$

Note the appearance of $S(t)$ on both sides of (3.20), which means that we have an equation with $S(t)$ as unknown. Dividing by dt leads to

$$\frac{dS(t)}{dt} = \left(\mu + \frac{1}{2}\sigma^2 + \sigma \frac{dB(t)}{dt}\right) S(t),$$

which we recognize as an ordinary differential equation with a non-standard term involving the time derivative of Brownian motion. We have earlier pointed out that this derivative does not exist, so the only way to make sense of (3.20) is as an integral equation. Equation (3.20) is the differential form of the integral equation

$$S(t) = S(0) + \int_0^t \left(\mu + \frac{1}{2}\sigma^2\right) S(u)\, du + \int_0^t \sigma S(u)\, dB(u), \tag{3.21}$$

3.3 Geometric Brownian Motion

and is an example of a *stochastic differential equation*. In the derivation of (3.20) we started out with (3.19). Hence, we know the solution of (3.20), or more precisely, (3.21).

If we instead define geometric Brownian motion as the solution of the stochastic differential equation

$$dS(t) = \alpha S(t)\,dt + \sigma S(t)\,dB(t), \tag{3.22}$$

a calculation similar to the above will reveal

$$S(t) = S(0)\exp\left(\left(\alpha - \frac{1}{2}\sigma^2\right)t + \sigma B(t)\right).$$

We see that a correction term $\sigma^2/2$ is introduced in the drift. It is important to note the difference in geometric Brownian motion defined as (3.19) and $S(t)$ defined as the solution of the stochastic differerential equation (3.22). When we are going to look at pricing and hedging of derivatives in Chap. 4 the differential form (3.22) is particularly useful since we will encounter situations where it is necessary to use Itô's formula for expressions of the type $C(t, S(t))$ for a certain function C.

Many options have several underlying stocks rather than just one. A multi-dimensional stock price model is therefore required, and we will here extend geometric Brownian motion for this purpose. We would like to have a price dynamics which models correlation *among* stocks at the same time as each stock follows a geometric Brownian motion. Assume that we want to describe the price evolution of n stocks with prices denoted by $S_1(t), \ldots, S_n(t)$. Let $B_1(t), B_2(t), \ldots, B_m(t)$ be m independent Brownian motions. Define the price dynamics of stock i to be the solution of the stochastic differential equation

$$dS_i(t) = \alpha_i S_i(t)\,dt + S_i(t)\sum_{j=1}^{m}\sigma_{ij}\,dB_j(t), \quad i = 1, \ldots, n, \tag{3.23}$$

which is called a *multi-dimensional* geometric Brownian motion. The reader is asked in Exercise 3.13 to demonstrate that the solution of (3.23) is

$$S_i(t) = S_i(0)\exp\left(\left(\alpha_i - \frac{1}{2}\sum_{j=1}^{m}\sigma_{ij}^2\right)t + \sum_{j=1}^{m}\sigma_{ij}B_j(t)\right), \tag{3.24}$$

for $i = 1, \ldots, n$. The number of independent Brownian motions m can be less than, equal to or greater than the number of stocks n. The volatility parameters σ_{ij} describe the correlation among the *logreturns* of the stocks. The reader is asked to calculate the correlations in Exercise 3.14.

The conclusion from this section is that geometric Brownian motion has two representations. Either we can define it as the stochastic process solving the stochastic differential equation (3.22) ((3.23) in the multi-dimensional case), or as the stochastic process (2.1). The reader is challenged to find the corresponding representation in multi-dimensions to (2.1). Both ways of defining geometric Brownian motion are equivalent however, we shall prefer the differential version when we move on to option pricing in Chap. 4.

3.4 Conditional Expectation and Martingales

The so-called *martingale processes*, or simply *martingales*, constitute an important class of stochastic processes. In mathematical finance they are one of the main building blocks for deriving option prices and hedging strategies. Stochastic analysis contains many different types of martingales, but we will focus on martingales *with respect to Brownian motion* only.

Expressed in words, a stochastic process $M(t)$ is a martingale if the best prediction of its state at time t, conditioned on the path of Brownian motion up to a previous time $s < t$, is the state at time s, $M(s)$. We make this property precise, and then continue to explain the different components of the definition.

Definition 3.10. *A stochastic process $M(t)$ is called a martingale if it is adapted and*

$$\mathbb{E}\left[M(t) \mid \mathcal{F}_s\right] = M(s), \qquad (3.25)$$

for every $0 \leq s \leq t < \infty$.

In order to fully understand the definition of a martingale, we need to grasp the concept of *conditional expectation*. On the left-hand side of (3.25) we are taking the expected value of $M(t)$ *conditioned* on all the information Brownian motion can give us up to time s. This information, which is encapsulated in the notation $\mathcal{F}_s$, has to be considered as the *potential* information being revealed to us as time progresses. Remark that if we know the path of Brownian motion from time 0 up to time s, we also know the evolution of the stock price up to time s (see (2.1)). So, when the path of Brownian motion is revealed, it is really the stock prices that become known to us. The information we condition on is potential because at time 0 we *do not* know the stock prices up to time s. What we *do* know is that they are generated by the stochastic process $B(u), 0 \leq u \leq s$. This is the information that we condition on in (3.25). The reader should note that this generalizes conditioning on events.

We have already met the notation $\mathcal{F}_s$ in connection with adaptedness of stochastic processes in Sect. 3.1. As we indicated, $\mathcal{F}_s$ denotes the collection of *all* the potential information from Brownian motion up to time s, that is, the collection of all potential price paths of the stock up to time s. In a sense that we will try to make clear, the expectation in (3.25) is conditioned on a whole collection of events. In fact, $\mathcal{F}_s$ is a family of events $A \subset \Omega$, where events A are derived from the states of Brownian motion. The events $A \in \mathcal{F}_s$ are of the form

3.4 Conditional Expectation and Martingales

$$A = \{\omega \in \Omega \mid B(s_1,\omega) \in H_1, B(s_2,\omega) \in H_2, \ldots, B(s_n,\omega) \in H_n,$$
$$0 \leq s_1 < \ldots s_n \leq s\},$$

for all possible natural numbers n and any subsets $H_1, \ldots H_n$ of the real line.[6] For instance, if $s = 10$ one event could be described by $n = 1$, $s_1 = 5$ and $H = (-1,1)$, the open interval between -1 and 1. The set A would in this case be the event that Brownian motion at a time 5 lies between -1 and 1. Another choice of A could be $n = 2$, $s_1 = 30, s_2 = 75$, $H_1 = (-2,4)$ and $H_2 = (2,8)$. Then we describe the event that $B(30) \in (-2,4)$ and $B(75) \in (2,8)$. In Fig. 3.1 we have plotted two paths of a Brownian motion where we see that the path for ω_2 passes both "hurdles" given by H_1 and H_2, while for ω_1 the path misses H_1. This means that $\omega_2 \in A$, while $\omega_1 \notin A$.

We can make the sets H more complicated than simply an open interval, for instance, a union of several intervals or intervals of the form $H = (a, \infty)$ or $H = (-\infty, b)$, where a, b are real numbers. We see that traversing through different specifications of A we single out the potential states of Brownian motion (that is, the potential stock prices) in the time interval 0 to s. Hence, if we collect all of these sets we end up with a complete description of all eventualities that can happen to Brownian motion in this time interval. This is what $\mathcal{F}_s$ constitutes.

With the collection of events $\mathcal{F}_s$ at hand, we define what is meant by $\mathbb{E}[Z|\mathcal{F}_s]$ for a random variable Z.

Definition 3.11. *Assume that Z is a random variable. Then the conditional expectation $\mathbb{E}[Z|\mathcal{F}_s]$ is defined as the $\mathcal{F}_s$-adapted random variable X satisfying*

$$\mathbb{E}[1_A X] = \mathbb{E}[1_A Z], \quad \text{for all } A \in \mathcal{F}_s. \tag{3.26}$$

One can show that there exists only one $\mathcal{F}_s$-adapted random variable X which satisfies (3.26). Hence, the conditional expectation is unique.

In order for the conditional expectation to exist, we need to impose a moment condition on Z: one can only define the conditional expectation of random variables Z for which $\mathbb{E}[|Z|] < \infty$. From Jensen's inequality (see [47, Thm. 19, p. 12]) it holds that

$$|\mathbb{E}[Z \mid \mathcal{F}_s]| \leq \mathbb{E}[|Z| \mid \mathcal{F}_s] \leq \mathbb{E}[|Z|], \tag{3.27}$$

---

[6] Unfortunately, this cannot be used as a rigorous mathematical definition of $\mathcal{F}_s$. For readers familiar with measure theory, $\mathcal{F}_s$ has to be defined as the smallest σ-algebra containing all the sets on the form A, that is, the smallest σ-algebra generated by Brownian motion up to time s. The subsets H_i must be Borel sets of the real line. In fact, $\mathcal{F}_s$ becomes a family of σ-algebras parametrized by time s. Our definition, or rather heuristic introduction of $\mathcal{F}_s$ is an attempt to describe this collection of sets without getting drowned in mathematical technicalities, and the highly skilled reader must therefore excuse our lack of rigour at this point.

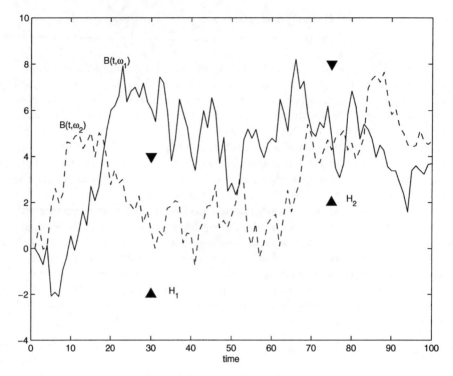

Fig. 3.1. Two paths of Brownian motion and two sets $H_1 = (-2, 4)$ and $H_2 = (2, 8)$

which shows that the conditional expectation is finite under this moment condition.

We now present some useful properties for the conditional expectation. First, conditional expectation is linear in the sense that

$$\mathbb{E}\left[aZ_1 + bZ_2 \,|\, \mathcal{F}_s\right] = a\mathbb{E}\left[Z \,|\, \mathcal{F}_s\right] + b\mathbb{E}\left[Z_2 \,|\, \mathcal{F}_s\right],$$

for two arbitrary constants a and b.

Next, notice that since $B(0) = 0$ by definition, $\mathcal{F}_0$ can only consist of two events $A = \Omega$ and $A = \emptyset$, the empty set (see Exercise 3.16). Define $X = \mathbb{E}\left[Z\right]$, which is $\mathcal{F}_0$-adapted since X is simply a constant. Observe that

$$\mathbb{E}\left[1_\Omega X\right] = \mathbb{E}\left[1_\Omega Z\right], \quad \mathbb{E}\left[1_\emptyset X\right] = 0 = \mathbb{E}\left[1_\emptyset Z\right],$$

which proves that $X = \mathbb{E}\left[Z \,|\, \mathcal{F}_0\right]$, or, $\mathbb{E}\left[Z \,|\, \mathcal{F}_0\right] = \mathbb{E}\left[Z\right]$. Hence, the conditional expectation with respect to $\mathcal{F}_0$ is simply the expectation itself, $\mathbb{E}\left[Z\right]$. When $M(t)$ is a martingale, this property entails

$$M(0) = \mathbb{E}\left[M(t) \,|\, \mathcal{F}_0\right] = \mathbb{E}[M(t)].$$

A necessary condition for a process to possess the martingale property is therefore constant expectation.

3.4 Conditional Expectation and Martingales

Another useful property is related to $\mathcal{F}_s$-adapted random variables. If Y is an $\mathcal{F}_s$-adapted random variable, it holds that

$$\mathbb{E}[YZ \mid \mathcal{F}_s] = Y\mathbb{E}[Z \mid \mathcal{F}_s]. \tag{3.28}$$

In Exercise 3.17 we prove that the conditional expectation of a constant is simply equal to the constant itself. Hence, choosing $Z = 1$, from (3.28) we obtain

$$\mathbb{E}[Y \mid \mathcal{F}_s] = Y,$$

whenever Y is $\mathcal{F}_s$-adapted.

Finally we introduce a very useful property that we recognize from conditional expectation on events. If we let $A = \Omega$ in the definition of conditional expectation, we get *the law of double expectation*,

$$\mathbb{E}\left[\mathbb{E}[Z \mid \mathcal{F}_s]\right] = \mathbb{E}[Z]. \tag{3.29}$$

Properties (3.28) and (3.29) will be used to calculate option prices in the subsequent chapter.

With some effort it is possible to demonstrate that all stochastic processes of the form

$$Z(t) = Z(0) + \int_0^t X(s) \, dB(s),$$

are martingales when $Z(0)$ is a constant. A central result for martingales is the so-called *martingale representation theorem*.

Theorem 3.12. *If $M(t)$ is a martingale, there exists an Itô integrable process X_s such that*

$$M(t) = M(0) + \int_0^t X(s) \, dB(s).$$

A consequence of this theorem is that we can *define* martingales as processes of the form $M(t) = M(0) + \int_0^t X(s) \, dB(s)$. We have chosen not to introduce martingales in this fashion, but rather present for the reader the concept of conditional expectation which yields the true martingale property. Having said this, we point out that later we will manipulate the conditional expectation and martingales to derive in a very simple way prices for options.

Exercises

3.1 Let $X(s)$ be an adapted process. Show that for an arbitrary partition $\{s_1, \ldots, s_n\}$ of the interval $[0, t]$ we have

$$\mathbb{E}\left[\left(\sum_{i=1}^n X(s_i)(B(s_{i+1}) - B(s_i))\right)^2\right] = \sum_{i=1}^n \mathbb{E}\left[X(s_i)^2\right](s_{i+1} - s_i).$$

3.2 Show that

$$\mathbb{E}\left[\int_0^t X(s)\,dB(s)\right] = 0, \quad \text{Var}\left(\int_0^t X(s)\,dB(s)\right) = \int_0^t \mathbb{E}\left[X(s)^2\right]\,ds.$$

3.3 Suppose that $X(s)$ and $Y(s)$ are two Itô integrable processes and a, b are two arbitrary constants. Prove that $aX(s) + bY(s)$ is Itô integrable. Why is $X(s) := 1$ an Itô integrable process?

3.4 Show that

$$\mathbb{E}\left[\left(\sum_{i=1}^{n-1} f''(B(s_i))\left\{(B(s_{i+1}) - B(s_i))^2 - (s_{i+1} - s_i)\right\}\right)^2\right] \longrightarrow 0,$$

when $n \to \infty$ and $\{s_i\}_{i=1}^n$ is a partition of the interval $[0, t]$.

3.5 Prove that if f'' is a continuous function, then the stochastic process $f''(B(t))$ has bounded paths on $[0, t]$. Demonstrate that

$$\mathbb{E}\left[\left(\int_0^t f''(B(s))\,ds\right)^2\right] < \infty,$$

when $\mathbb{E}\left[\int_0^t f''(B(s))^2\,ds\right] < \infty$. To show this, you need the Cauchy–Schwarz inequality, which in our context says that

$$\int_a^b g(s)h(s)\,ds \le \left(\int_a^b g^2(s)\,ds\right)^{1/2} \left(\int_a^b h^2(s)\,ds\right)^{1/2}, \quad (3.30)$$

for two functions g and h. In [25] you will find a more general version of this useful inequality.

3.6 If $X(t)$ is the semimartingale

$$X(t) = x + \int_0^t Y(s)\,dB(s) + \int_0^t Z(s)\,ds,$$

where $Y(s)$ and $Z(s)$ are Itô integrable processes, then $\mathbb{E}\left[X^2(t)\right] < \infty$. You may find the rough estimate $(a+b)^2 \le 2a^2 + 2b^2$ and the Cauchy–Schwarz inequality (3.30) useful in this exercise.

3.7 Verify that all integral terms in Thm. 3.7 exist and have finite second moment under conditions (3.14) and (3.15).

3.8 Verify conditions (3.14) and (3.15) in Example 3.8.

3.9 Let $B(0) = 0$ and $h(s)$ be a real-valued function which is differentiable and such that $\int_0^t h^2(s)\,ds < \infty$. Show that $h(s)$ is Itô integrable and use Itô's formula to prove the identity

$$\int_0^t h(s)\,dB(s) = h(t)B(t) - \int_0^t h'(s)B(s)\,ds.$$

3.4 Conditional Expectation and Martingales

3.10 If X is a random variable, we define (whenever it exists) $f(\theta) := \mathbb{E}\left[e^{\theta X}\right]$ for $\theta \in \mathbb{R}$ to be the *moment generating function* of X.
a) Prove that $f^{(n)}(0) = \mathbb{E}[X^n]$, where $f^{(n)}$ is the nth derivative of f.
b) Assume $X \sim \mathcal{N}(0, a^2)$. Prove that

$$\mathbb{E}\left[e^{\theta X}\right] = e^{\frac{1}{2}\theta^2 a^2}. \tag{3.31}$$

In fact, the opposite holds. If X is a random variable which has the moment generating function (3.31), then $X \sim \mathcal{N}(0, a^2)$. Prove this using the Fourier transform.
c) Show that

$$\int_0^t h(s)\,\mathrm{d}B(s) \sim \mathcal{N}\left(0, \int_0^t h^2(s)\,\mathrm{d}s\right),$$

where we assume h to be a real-valued function with $\int_0^t h^2(s)\,\mathrm{d}s < \infty$.
Hint: use Itô's formula to find $\mathbb{E}[\exp(\theta \int_0^t h(s)\,\mathrm{d}B(s))]$.

3.11 Let $S(t) = S(0)\exp(\mu t + \sigma B(t))$. Use Itô's formula with $X(t) = \sigma B(t)$ and $X(t) = \mu t + \sigma B(t)$ to calculate $\mathrm{d}S(t)$. Find $S(t)$ when

$$\mathrm{d}S(t) = \alpha S(t)\,\mathrm{d}t + \sigma S(t)\,\mathrm{d}B(t).$$

3.12 Let $Y(t) = \exp(B_1(t) + B_2(t))$ for two independent Brownian motions $B_1(s)$ and $B_2(s)$. Use the multi-dimensional Itô formula to find $\mathrm{d}Y(t)$.

3.13 Demonstrate using the multi-dimensional Itô formula that the solution of (3.23) is

$$S_i(t) = S_i(0)\exp\left(\left(\alpha_i - \frac{1}{2}\sum_{j=1}^m \sigma_{ij}^2\right)t + \sum_{j=1}^m \sigma_{ij}B_j(t)\right), \quad i=1,\ldots,n.$$

3.14 Let

$$\mathrm{d}S_1(t) = \alpha_1 S_1(t)\,\mathrm{d}t + S_1(t)\left(\sigma_{11}\,\mathrm{d}B_1(t) + \sigma_{12}\,\mathrm{d}B_2(t)\right)$$
$$\mathrm{d}S_2(t) = \alpha_2 S_2(t)\,\mathrm{d}t + S_2(t)\left(\sigma_{21}\,\mathrm{d}B_1(t) + \sigma_{22}\,\mathrm{d}B_2(t)\right),$$

where $B_1(t)$ and $B_2(t)$ are two independent Brownian motions. Calculate the correlation between the logreturns of $S_1(t)$ and $S_2(t)$, that is, find the correlation between $X(t) = \ln(S_1(t)/S_1(t-1))$ and $Y(t) = \ln(S_2(t)/S_2(t-1))$.

3.15 Let $H_1 = (0, \infty)$ and $H_2 = (-\infty, 0)$.
a) Define

$$A = \{\omega \in \Omega \mid B(0.5, \omega) \in H_1, B(2, \omega) \in H_2\}.$$

Draw some paths of Brownian motion that are in A and some paths that are not in A. Describe A in terms of the stock price $S(t) = S(0)\exp(\mu t + \sigma B(t))$.

b) Let $A = \{\omega \in \Omega \,|\, B(0.5, \omega) \in H_1\}$. What is the probability that a path of Brownian motion will be in A?

c) Let $0 < a < b < \infty$ and define $A = \{\omega \in \Omega \,|\, S(1, \omega) \in (a, b)\}$ where $S(t)$ is as in a) above. Show that $A \in \mathcal{F}_1$.

3.16 Argue that the only two events in $\mathcal{F}_0$ are either Ω or the empty set $\emptyset$.

3.17 Let k be a constant. Show that $\mathbb{E}\left[k \,|\, \mathcal{F}_s\right] = k$.

4 Pricing and Hedging of Contingent Claims

The goal of this chapter is to derive fair prices of derivatives contracts, that is, financial contracts that depend on an underlying stock. Furthermore, we shall discuss how one can hedge the risk associated with a position in a derivatives contract. The price dynamics of the underlying stock is assumed to be modelled by geometric Brownian motion, that is, the Black & Scholes model. One of the highlights of the chapter is the famous Black & Scholes option pricing formula, which states the fair value of a call option. Rather than restricting our attention to call and put options, we will consider a very general class of derivatives contracts called *contingent claims*. The reason for going to such generality is that we would like to include popular derivatives like Asian options, barrier options, chooser options etc. Unfortunately, American-type options will not fit into our framework.

A contingent claim is defined as follows:

Definition 4.1. *A contingent T-claim is a financial contract that pays the holder a random amount X at time T. The random variable X is $\mathcal{F}_T$-adapted, and T is called the* exercise time *of the contingent claim.*

Recall from the considerations about $\mathcal{F}_T$ in Sect. 3.4 that it contains all information about $B(s)$ up to time T. Since we have assumed the Black & Scholes model for the stock price, the potential information from $B(s)$ up to time T coincides with all potential price paths of the stock up to time T. A contingent T-claim is therefore a random variable that depends on the stock price at one or more time instances between 0 and T. The random variable X is the payoff at the time of exercise T. When there is no chance of confusion, we shall from now on call X a contingent claim, or simply a claim, and skip the reference to T. Later, we shall impose moment conditions on the random variable X.

Observe that all contracts with payoff $f(S(T))$, where f is some function, are contingent claims. Let $X = f(S(T))$, and we see that X is dependent on $S(T)$ and thus $B(T)$, which implies that X is $\mathcal{F}_T$-adapted. By choosing $f(x) = \max(x - K, 0)$ or $f(x) = \max(K - x, 0)$, call and put options are included. A barrier option is an option which is knocked-out (that is, pays zero) if the stock price breaks a prescribed barrier during the lifetime of the option. Otherwise the option pays the same as a call or put option. One

54   4 Pricing and Hedging of Contingent Claims

example of a barrier option is a call option with strike price K being knocked-out if the underlying stock price is above the barrier $\beta > K$ at some time during 0 and T. The payoff from this option is

$$X = 1_{\{S(t) < \beta, \forall 0 \leq t \leq T\}} \max\left(S(T) - K, 0\right),$$

which we see depends on $S(t)$, and thus $B(t)$, for all $t \in [0, T]$. Hence, the barrier option is a contingent claim. An Asian option is an example of an average option. A typical Asian option can be a contract that pays the holder in cash the difference between the average stock price over the interval $[0, T]$ and a strike price K, whenever this difference is positive, and zero otherwise. The Asian option is thus a call option on the average stock price and has payoff function

$$X = \max\left(\left(\frac{1}{T}\int_0^T S(t)\,\mathrm{d}t - K\right), 0\right),$$

which also fits into the definition of a contingent claim.

The problem of pricing and hedging contingent claims is solved using the *martingale technique*, which exploits the martingale property of certain stochastic processes connected to the contingent claim. If we limit our considerations to the special class of contingent claims having payoff $X = f(S(T))$ for some function f, one can instead use Itô's formula to derive a partial differential equation for the price and hedge of the claim. By solving the partial differential equation, we end up with the Black & Scholes formula for the price of a call option. We begin the chapter with these derivations, and then continue with the martingale technique for general claims.[1] Since we assume that the underlying stock follows a geometric Brownian motion, the theory of pricing and hedging of contingent claims is known as the Black & Scholes option pricing theory. However, to motivate the main components in this theory, let us start with an example from one-period markets.

4.1 Motivation from One-Period Markets

To motivate some of the ideas of the Black & Scholes option theory, we consider the problem of pricing and hedging a contingent claim in a one-period market. In a one-period market the derivations become particularly simple, enabling us to focus on the important steps and ideas.

Suppose our market consists of a stock and a bond for which we model prices just at some future time T. Assume the bond has price 1 at time 0 and

---

[1] The reader who is interested to go into deeper detail on option theory is advised to read the textbooks by Bjørk [7], Duffie [20], Karatzas and Shreve [34], Musiela and Rutkowski [40] or Øksendal [41]. All these books introduce rigorously the option pricing theory, and have been a rich source of inspiration for the author.

4.1 Motivation from One-Period Markets

the stock has price s_0. The stock price is assumed to have two outcomes at time T, so we let $\Omega = \{\omega_1, \omega_2\}$ and put $S(T, \omega_1) = s_1$ and $S(T, \omega_2) = s_2$,

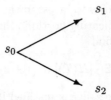

For convenience, we suppose $s_1 > s_2$, and denote the probability of an increase in stock price by $p := \mathcal{P}(\omega_1) \in (0,1)$. Assume the rate of return from the bond is r, which gives a bond value $1+r$ at time T. We start by describing possible investments in this market.

If we buy a stocks and b bonds today, our portfolio will at time T have the value
$$H(T, \omega_1) = as_1 + b(1+r),$$
with probability p, or
$$H(T, \omega_2) = as_2 + b(1+r),$$
with probability $1 - p$. Our investment will cost us $H(0) = as_0 + b$.

A question that will become relevant later is whether one can find a constant q such that
$$H(0) = (1+r)^{-1} \left(qH(T, \omega_1) + (1-q)H(T, \omega_2) \right).$$

The answer is yes, and the reader can easily check that $q = (s_0(1+r) - s_2)/(s_1 - s_2)$. If $q \in (0,1)$, we can write
$$H(0) = (1+r)^{-1} \mathbb{E}_{\mathcal{Q}}[H(T)], \tag{4.1}$$
where the probability $\mathcal{Q}$ on Ω is defined by $\mathcal{Q}(\omega_1) = q$ and $\mathcal{Q}(\omega_2) = 1 - q$. Note that $q \neq p$ in general, and is therefore *not* the real probability for an increase in the stock price. We conclude that for every portfolio the value today, $H(0)$ is the expected discounted future value, where the expectation is taken with respect to the probability $\mathcal{Q}$. Investing $a = 1$ and $b = 0$, we can show that
$$s_0 = (1+r)^{-1} \mathbb{E}_{\mathcal{Q}}[S(T)]. \tag{4.2}$$

This motivates the name *risk-neutral probability* for $\mathcal{Q}$, since the expected rate of return from an investment in the stock is the same as for the risk-free bond, when working with the probability $\mathcal{Q}$.

What does this have to do with options? The connection is pretty straighforward. Assume we want to buy a contingent claim X (for instance a call option on the stock) and wonder if the seller suggests a reasonable price. To sort this out, we calculate the a and b such that the portfolio perfectly matches the claim at time T, that is, $H(T) = X$. The price of the claim, $P(0)$ should now be the same as it costs to invest in the portfolio, $H(0)$. This portfolio is called the hedging or replicating portfolio (hence the notation H). From the considerations above, we find

$$P(0) = H(0) = (1+r)^{-1}\mathbb{E}_Q\left[H(T)\right] = (1+r)^{-1}\mathbb{E}_Q\left[X\right].$$

In conclusion, the price asked by the seller is justified if it coincides with

$$P(0) = (1+r)^{-1}\mathbb{E}_Q\left[X\right]. \tag{4.3}$$

In the continuous-time model of Black & Scholes the price will also be represented as the discounted expected payoff under a risk-neutral probability.

Two questions remain to be answered: does there exist a portfolio that can perfectly replicate the contingent claim X, and why is it so that the price of the claim *should* be equal to the investment in the replicating portfolio? The first question is connected with the notion of *completeness* of the market, that is, whether you can replicate claims or not. The second question relates to absence of *arbitrage*.

Let us answer the latter question first. We claim that all prices other than $P(0)$ will allow for arbitrage opportunities. For example, assume that the market trades in the claim for a price $\widetilde{P} < P(0)$. A rational investment would then be to exploit this bargain by *buying* N claims. The money you need for this investment, $N\widetilde{P}$, you take from short-selling N replicating portfolios, providing you with $NH(0)$. Since $H(0) = P(0)$ and $P(0) > \widetilde{P}$, you have a surplus of $N(H(0) - \widetilde{P})$ that you can invest in whatever you like. Let us say we choose to put this money in bonds. We can buy exactly $N(H(0) - \widetilde{P})$ bonds. All we do next is wait until time T is reached, when we exercise our claims and earn NX. At the same time, we need to settle our short position in the replicating portfolios, costing us $NH(T) = NX$. Our net position is thus zero, using the claims to cover our debts. The bond position has value $N(H(0) - \widetilde{P})(1+r)$, which is a profit with certainty. In fact by appropriately choosing N we can make an arbitrary large profit without taking any risk, which is known as an *arbitrage opportunity*. If $\widetilde{P} > P(0)$, we *issue* N claims and sell them in the market. For the money we receive we buy N replicating portfolios. In Exercise 4.1 the reader is asked to elaborate the details for this case and show that it also gives an arbitrage opportunity. We conclude that if the market is assumed not to allow for any arbitrage opportunities, the claim must have a price equal to the cost of investing in the replicating portfolio.

Regarding the question of replicating the claim, this is a simple calculation in linear algebra. First, denote $X(\omega_1) = x_1$ and $X(\omega_2) = x_2$. In order for $H(T) = X$, we must find a and b such that

$$as_1 + b(1+r) = x_1,$$

4.1 Motivation from One-Period Markets

$$as_2 + b(1+r) = x_2.$$

This 2×2 system of linear equations has a unique solution

$$a = \frac{x_1 - x_2}{s_1 - s_2}, \quad b = (1+r)^{-1}\left(x_1 - s_1\frac{x_1 - x_2}{s_1 - s_2}\right).$$

The solution a and b will describe how to invest in order to replicate perfectly the claim. It is not always possible to replicate claims (see, e.g. Exercise 4.2), and such markets are called *incomplete*. There can be many reasons for incompleteness, which we discuss in Sect. 4.9. We shall mainly study markets which are complete. Note, however, that from a practical point of view this may not be the interesting case, since the claim does not have any purpose in such a market. One can invest in the replicating portfolio instead, thus making the claim redundant. As we know, the option markets exist, indicating that these finanical instruments add new investment opportunities, and they can only do so because the market is incomplete. We postpone any further discussion of this to Sect. 4.9.

Note that the price $P(0)$ and the replicating portfolio are independent of the probability p that the stock price will increase. To appreciate this peculiar fact, consider the following example. Let X be a call option with strike $K = 100$, and assume that $r = 0$ (no interest is paid on the bond). Let furthermore $s_0 = 100$, $s_1 = 120$ with probability p, and $s_2 = 95$ with probability $1-p$. From the calculations above we find the replicating portfolio to be $a = 0.8$ and $b = -76$, that is, short-sell 76 bonds and buy 0.8 shares of the stock. This will cost $H(0) = 4$, hence the price of the call is €4. If the probability that the stock will be 120 is 99% at time T ($p = 0.99$), we will be 99% certain to receive €20 from an investment of €4. If the probability that the stock increases is 1% ($p = 0.01$), we find that we will lose the whole investment of €4 with 99% probability. For the first time in this book we meet the fact that the option price is only dependent on the variation in possible prices of the underlying stock, and not on the expected return (which is connected to p).

We summarize our findings from this section.

- In a market without arbitrage opportunities all claims that can be replicated will have a price equal to the cost of investing in the replicating portfolio.
- This price can be expressed using the risk-neutral probability $\mathcal{Q}$ as $P(0) = (1+r)^{-1}\mathbb{E}_\mathcal{Q}[X]$.

The next step is to generalize these considerations to a market where the stock is modelled by geometric Brownian motion. The reason for doing this is of course that the one-period market model is very unrealistic, and to have confidence in the derived prices we need to base our findings on models which are close to real market behaviour.

4.2 The Black & Scholes Market and Arbitrage

We consider a financial market consisting of a stock (risky investment), a bond (risk-free investment) and a contingent claim. The price process of the stock is modelled by a geometric Brownian motion,

$$dS(t) = \alpha S(t)\,dt + \sigma S(t)\,dB(t), \tag{4.4}$$

while the price dynamics of the bond takes the form

$$dR(t) = rR(t)\,dt. \tag{4.5}$$

We normalize the price of the bond to be initially $R(0) = 1$. The constant r is the *continuously compounding* rate of return.[2] The solution of the bond dynamics is $R(t) = \exp(rt)$ (show this yourself by simple differentiation), and the link to the actuarial rate of return r_a is through the formula $\exp(rt) = (1+r_a)^t$, or, equivalently, $r_a = \exp(r) - 1$. Finally, we let the price dynamics of the contingent claim be an adapted stochastic process denoted by $P(t)$. The purpose of this chapter is to derive the dynamics of this price process, and it will turn out that the process has the adaptedness property. In fact, if the price process of the claim were not to be adapted, it would contain information about future values of the stock price. In a liquid market, there are no reasons why a claim should hold more information about future stock price than the stock itself. Hence, imposing the assumption of adaptedness is quite natural at this stage.

An investor in our Black & Scholes market can form portfolios from the three investment alternatives. Let $a(t)$ be the number of stocks, $b(t)$ the number of bonds and $c(t)$ the number of claims in such a portfolio at time t, which are all assumed to be adapted stochastic processes. We call (a,b,c) the *portfolio strategy*, and the value at time t is

$$V(t) = a(t)S(t) + b(t)R(t) + c(t)P(t). \tag{4.6}$$

The value process $V(t)$ becomes an adapted stochastic process. An investor will not have access to more information than is revealed through the market-quoted prices, which implies that the strategy cannot be based on information about future stock prices. The portfolio strategies can therefore only be adapted. We shall pay particular attention to portfolios which are *self-financing*. In words, the self-financing property means that the investor is not withdrawing any gains from the portfolio for consumption, nor investing additional funds. She starts with an initial investment, and from there on all gains or losses in portfolio value come from price increases or decreases in the stock, bond or claim. Furthermore, the property tells us that if the investor

---

[2] One usually lets r be the rate of return from Treasury bills, that is, non-coupon bearing bonds secured by the Government with time to maturity usually up to one year.

4.2 The Black & Scholes Market and Arbitrage

wants to increase the stock position, say, the funding for this must come from selling bonds and claims. Mathematically, one defines self-financing as follows.

Definition 4.2. *A portfolio strategy (a, b, c) is called self-financing if*

$$\mathrm{d}V(t) = a(t)\,\mathrm{d}S(t) + b(t)\,\mathrm{d}R(t) + c(t)\,\mathrm{d}P(t). \tag{4.7}$$

Recall that the differential form of a stochastic process is just a simplified notation, and the self-financing property manifests itself as

$$V(t) = V(0) + \int_0^t a(s)\,\mathrm{d}S(s) + \int_0^t b(s)\,\mathrm{d}R(s) + \int_0^t c(s)\,\mathrm{d}P(s).$$

Since we do not know whether or not $P(t)$ is a semimartingale, we cannot say what should be the interpretation of the stochastic integral with respect to $P(s)$. However, the heuristic interpretation of (4.7) is clear. We can think of $\mathrm{d}V$ as the change in portfolio value, and if the portfolio is self-financing, this change is equal to the stock position times the change of stock price added to the bond position times the change of bond, plus the claim position times the change of claim price. It will later turn out that $P(t)$ is indeed a semimartingale.

Arbitrage is the possibility of earning money from a zero investment without taking any risk. For instance, if there is a way to short-sell bonds to finance a purchase of stocks and claims which yields a sure profit, the market is pricing the different instruments so that arbitrage is possible. In ideal markets, such opportunities should not exist, simply because the investors will see this opportunity and try to exploit it by competing on prices. This would eventually lead to an equlibrium price in a liquid market, where arbitrage is not possible. We define what is understood by an arbitrage opportunity:

Definition 4.3. *A self-financing portfolio strategy is called an* arbitrage *opportunity if $V(0) \leq 0$, $V(T) \geq 0$ and $\mathbb{E}\left[V(T)\right] > 0$.*

A liquid market does not allow for any arbitrage. This will be one of the main properties on the way to deriving a price dynamics for the claim.

Our objective is to find self-financing portfolios which can hedge, or replicate the claim with an investment in the stock and bond only. The value of such portfolios will be denoted by $H(t)$, and for a given strategy (a^H, b^H) we have

$$H(t) = a^H(t)S(t) + b^H(t)R(t).$$

In order for H to be a hedging portfolio, it must hold that $H(T) = X$, that is, the value of the hedge must be the same as the payoff of the claim at the time of exercise. If all claims in the market can be hedged, we say that the market is *complete*. Hedging portfolios will be used together with absence of arbitrage to find $P(t)$.

Unfortunately, there exist self-financing investment strategies (a, b) in the stock and bond which pay an arbitrary amount from a zero investment. These investment opportunities are called *doubling strategies*, and can be ruled out by imposing the following technical condition on the portfolio value. For each self-financing strategy (a, b), there exists a constant K such that $V(t) \geq K$ for all $t \in [0, T]$. For a concrete example of a doubling strategy and an explanation of the reasoning behind the name, see [20, Chap. 6D]. In our consideration we shall assume that this technical condition is always valid.

We end this section by stating the two main assumptions in the Black & Scholes market.

Assumptions: There are no arbitrage opportunities in the market and all claims can be hedged. Hence, we assume a *complete* and *arbitrage-free* market.

4.3 Pricing and Hedging of Contingent Claims $X = f(S(T))$

We consider a contingent claim with payoff at exercise time given as a function of the underlying stock, $X = f(S(T))$. We assume throughout that f is a function such that $\mathbb{E}\left[f(S(T))^2\right] < \infty$.

A claim which pays $X = f(S(T))$ at the exercise time T will have a price $P(t)$ at time t which is a function of the underlying stock price $S(t)$. We can therefore write $P(t) = C(t, S(t))$ for a function $C(t, x)$. The price we have to pay to purchase the claim today will be $P(0) = C(0, S(0))$. The objective of this section is to show that the function $C(t, x)$ can be derived as the solution of a partial differential equation. Furthermore, we will exhibit the relation between the price of the claim and a conditional expectation under a risk-neutral probability. The main result of this section is a complete description of the price and hedge of a claim, and as a special case we will obtain the famous Black & Scholes option pricing formula.

4.3.1 Derivation of the Black & Scholes Partial Differential Equation

Assume we can represent the price as $P(t) = C(t, S(t))$ for a function $C(t, x)$ which can be differentiated twice with respect to x and once with respect to t. Appealing to Itô's formula,[3] we get

$$dP(t) = \left\{ \frac{\partial C(t, S(t))}{\partial t} + \alpha \frac{\partial C(t, S(t))}{\partial x} S(t) + \frac{1}{2}\sigma^2 \frac{\partial^2 C(t, S(t))}{\partial x^2} S^2(t) \right\} dt$$
$$+ \sigma \frac{\partial C(t, S(t))}{\partial x} S(t) \, dB(t). \tag{4.8}$$

---

[3] We quietly assume the integrability conditions in Sect. 3.2.

4.3 Pricing and Hedging of Contingent Claims $X = f(S(T))$

Since we have assumed that the market is complete, all claims can be replicated. Hence, there exists a self-financing portfolio strategy (a^H, b^H) such that $H(T) = f(S(T))$. The no-arbitrage assumption forces $H(t) = P(t)$, otherwise we can construct self-financing portfolio strategies which become arbitrage opportunities. We shall discuss this in a little while, but for now we accept this fact and move on and use the self-financing property to obtain

$$\mathrm{d}P(t) = \mathrm{d}H(t) = a^H(t)\,\mathrm{d}S(t) + b^H(t)\,\mathrm{d}R(t)$$
$$= \left(\alpha a^H(t)S(t) + rb(t)R(t)\right)\mathrm{d}t + \sigma a^H(t)S(t)\,\mathrm{d}B(t). \quad (4.9)$$

Comparing the Itô integral terms in (4.8) and (4.9), we find that

$$a^H(t) = \frac{\partial C(t, S(t))}{\partial x}. \quad (4.10)$$

Further, the relation $C(t, S(t)) = P(t) = H(t) = a^H(t)S(t) + b^H(t)R(t)$ gives

$$b^H(t) = R^{-1}(t)\left(C(t, S(t)) - a^H(t)S(t)\right). \quad (4.11)$$

We see that the stochastic processes $a^H(t)$ and $b^H(t)$ are adapted, and by definition they will be self-financing. Hence, we have found the replicating strategy for the claim $X = f(S(T))$ as long as we know $C(t, x)$. Our next task is to derive an equation for C.

Inserting the expressions for $a^H(t)$ and $b^H(t)$ into (4.9), and then comparing the $\mathrm{d}t$-term with that in (4.8), we get

$$\frac{\partial C(t, S(t))}{\partial t} + \alpha S(t)\frac{\partial C(t, S(t))}{\partial x} + \frac{1}{2}\sigma^2 S^2(t)\frac{\partial^2 C(t, S(t))}{\partial x^2}$$
$$= \alpha S(t)\frac{\partial C(t, S(t))}{\partial x} + r\left(C(t, S(t)) - S(t)\frac{\partial C(t, S(t))}{\partial x}\right),$$

or,

$$\frac{\partial C(t, S(t))}{\partial t} + rS(t)\frac{\partial C(t, S(t))}{\partial x} + \frac{1}{2}\sigma^2 S^2(t)\frac{\partial C(t, S(t))}{\partial x^2} = rC(t, S(t)).$$

Hence, the function C has to solve the partial differential equation

$$\frac{\partial C(t, x)}{\partial t} + rx\frac{\partial C(t, x)}{\partial x} + \frac{1}{2}\sigma^2 x^2\frac{\partial C(t, x)}{\partial x^2} = rC(t, x), \quad (4.12)$$

for all $x \geq 0$. In addition, we have $C(T, x) = f(x)$ since $C(T, S(T)) = P(T) = f(S(T))$. Note that x varies along the positive real axis, while time is in the interval $[0, T]$. The equation (4.12) is called the Black & Scholes partial differential equation, and we conclude that the price of a claim $X = f(S(T))$ is $P(t) = C(t, S(t))$, where $C(t, x)$ is a solution of this. We say that the price $P(t)$ is *arbitrage-free*, because it is the only price dynamics that does not allow for any arbitrage opportunities in the market. Recall that we assumed an arbitrage-free market.

Before we solve (4.12), let us discuss why $P(t) \neq H(t)$ would lead to arbitrage opportunitites. Suppose that the market trades the claim for a price $\widetilde{P}(0) < H(0)$ at time 0, and that the claim thereafter follows an adapted stochastic process $\widetilde{P}(t)$. How can we exploit this mispricing in order to obtain riskless profit? Since the claim apparently is cheaper than it should be, we buy one and hold it in our portfolio until exercise, that is, let $c(t) = 1$. Then we enter a short position in the hedge. This will give us $H(0)$, which we partly use to buy the claim. The rest, $H(0) - \widetilde{P}(0) > 0$, we invest in the bond. Hence, our portfolio strategy becomes $(-a^H(t), -b^H(t) + H(0) - \widetilde{P}(0), 1)$, which is an adapted investment strategy. The portfolio has value $V(0) = 0$ and at time of exercise the value is

$$V(T) = -H(T) + (H(0) - \widetilde{P}(0))R(T) + \widetilde{P}(T)$$
$$= -X + (H(0) - \widetilde{P}(0))R(T) + X$$
$$= (H(0) - \widetilde{P}(0))R(T),$$

which is strictly positive. We need to check that our strategy is self-financing before we can conclude that we have an arbitrage opportunity. We write

$$V(t) = -H(t) + (\widetilde{P}(0) - H(0))R(t) + \widetilde{P}(t),$$

and observe from the self-financing property of $H(t)$ that

$$\mathrm{d}V(t) = -\mathrm{d}H(t) + (\widetilde{P}(0) - H(0))\,\mathrm{d}R(t) + \mathrm{d}\widetilde{P}(t)$$
$$= -a^H(t)\,\mathrm{d}S(t) + \left\{-b^H(t) + (\widetilde{P}(0) - H(0))\right\}\,\mathrm{d}R(t) + \mathrm{d}\widetilde{P}(t).$$

The strategy is indeed self-financing, and therefore an arbitrage opportunity. The reader is encouraged to work out the case $\widetilde{P}(t) > H(t)$ in Exercise 4.5.

The definition of $a^H(t)$ is the originator of what is known in the finance industry as *delta hedging*. Notice that $a^H(t) = \partial C(t, S(t))/\partial x$ has the alternative notation

$$a^H(t) = \frac{\partial P(t)}{\partial S(t)},$$

which says that the hedging position in the stock should match the sensitivity of the price with respect to the price of the underlying stock. This sensitivity has been assigned the Greek name *delta*, and is an important parameter for option traders. Note that $a^H(t)$ will vary continously with time, meaning that we should change our position at each moment. In practice such a strategy would become infinitely expensive due to transaction costs every time one wants to buy or sell stocks and bonds. This is a first indication of incompleteness of the market, since the only way to hedge the option is by using the practically impossible strategy $a^H(t)$. Later we shall discuss this problem and possible ways out of it. But for the time being, we continue with the belief that at least approximately it should be possible to delta-hedge (and indeed it is!).

4.3.2 Solution of the Black & Scholes Partial Differential Equation

The solution of (4.12) can be represented as the expectation of a random variable intimately connected to the payoff $X = f(S(T))$ of the claim. From this expression the Black & Scholes pricing formula for call options is simple to calculate. In Sect. 4.4 we establish the link between $\mathbb{E}_Q[X]$ and the solution $C(t,x)$, which means to define an appropriate probability Q. Before we explore these questions further, let us motivate ourselves with a simple example illuminating the ideas that our calculations are based on.

Consider Brownian motion starting at x, $B^x(t) := x + B(t)$, which is a normally distributed random variable with expectation x and variance t. Hence, for a function g we have

$$\mathbb{E}[g(B^x(t))] = \int_{-\infty}^{\infty} g(y) p(t, x-y) \, dy,$$

where $p(t,z) = (1/\sqrt{2\pi t}) \exp(-z^2/2t)$ is the probability density function of $B^x(t)$. The reader can check (see Exercise 4.6) that

$$\frac{\partial p(t,z)}{\partial t} = \frac{1}{2} \frac{\partial^2 p(t,z)}{\partial z^2}.$$

Defining the function $u(t,x) := \mathbb{E}[g(B^x(t))]$, a straightforward differentiation[4] with respect to t and x leads to

$$\frac{\partial u(t,x)}{\partial t} = \frac{1}{2} \frac{\partial^2 u(t,x)}{\partial x^2}. \tag{4.13}$$

Furthermore, we find that $u(0,x) = \mathbb{E}[g(x)] = g(x)$. The function $u(t,x)$ is a solution of a partial differential equation with initial value $g(x)$. The French mathematician, and the originator of mathematical finance, Bachelier [2] built up a theory around this equation. Bachelier preceded Einstein's classical work from 1905 on (4.13), which in physics is known as the *heat equation*, a partial differential equation modelling the diffusion of heat or movement of solid particles suspended in liquids.

Let us reverse time, and define the function $v(t,x) := u(T-t,x)$. Then $\partial v/\partial t = -\partial u/\partial t$ and $v(t,x)$ is the solution of the partial differential equation

$$\frac{\partial v(t,x)}{\partial t} + \frac{1}{2} \frac{\partial^2 v(t,x)}{\partial x^2} = 0, \tag{4.14}$$

with *terminal* condition $v(T,x) = u(0,x) = g(x)$. Recall from the definition of $u(t,x)$ that $v(t,x) = u(T-t,x) = \mathbb{E}[g(B^x(T-t))]$, where $B^x(T-t)$ is a normally distributed random variable with expectation x and variance

---

[4] In fact, this differentiation is not *that* straightforward. The mathematically careful reader will notice that certain conditions on g must be fullfilled in order to commute differentiation and integration. We drop such technicalities here.

$T - t$. Define Brownian motion which starts in x at time t as, for example, $B^{t,x}(s) := x + B(s) - B(t)$, $s \geq t$. It is easily seen that $B^{t,x}(T)$ also will be a normally distributed random variable with expectation x and variance $T - t$. Hence, we can write
$$v(t,x) = \mathbb{E}\left[g(B^{t,x}(T))\right],$$
as the solution of (4.14) with terminal condition $v(T,x) = g(x)$.

Equation (4.14) has some similarities with (4.12), and we can in fact write the solution $C(t,x)$ in an analogous way as $v(t,x)$. However, the process can no longer be Brownian motion, but will be a geometric Brownian motion corresponding to $S(t)$ except that α is substituted with r.

Theorem 4.4. *Let $Z^{t,x}(s)$, for $s \geq t$, be the stochastic process*
$$Z^{t,x}(s) = x + \int_t^s rZ^{t,x}(u)\,du + \int_t^s \sigma Z^{t,x}(u)\,dB(u). \tag{4.15}$$

Then
$$C(t,x) = e^{-r(T-t)}\mathbb{E}\left[f(Z^{t,x}(T))\right], \tag{4.16}$$

is the solution to the Black & Scholes partial differential equation (4.12).

It is not difficult to prove this result. An argument can be based on the derivations for Brownian motion, since the process $Z^{t,x}(s)$ is nothing but a geometric Brownian motion starting at x at time t,
$$Z^{t,x}(s) = x\exp\left(\left(r - \frac{1}{2}\sigma^2\right)(s-t) + \sigma(B(s) - B(t))\right).$$

Thus, $\ln(Z^{t,x}(T))$ is a normally distributed random variable with expectation $\ln x + (r - \sigma^2/2)(T-t)$ and variance $\sigma^2(T-t)$. The distribution of $Z^{t,x}(T)$ is therefore lognormal, and the considerations above can be adapted by a suitable choice of the function g. We elaborate the steps in Exercise 4.7. Note that the term $-rC(t,x)$ in (4.12) stems from the discounting by r.

The expression (4.16) gives us the price of the claim $X = f(S(T))$ along with the hedging strategy. We collect our findings in the following theorem.

Theorem 4.5. *The price $P(t)$ of the contingent claim $X = f(S(T))$ is given by*
$$P(t) = C(t, S(t)),$$

with a hedging strategy defined as
$$a^H(t) = \frac{\partial C(t, S(t))}{\partial x}, \quad b^H(t) = R^{-1}(t)\left(C(t, S(t)) - \frac{\partial C(t, S(t))}{\partial x}S(t)\right).$$

The function $C(t,x)$ is defined in (4.16).

A consequence of the theorem is the famous Black & Scholes formula for call options which we now present.

4.3.3 The Black & Scholes Formula for Call Options

Let the contingent claim be a call option on the stock with strike price K, that is, the claim has payoff function $X = \max(0, S(T) - K)$ at exercise time T. The function $f(x)$ becomes $f(x) = \max(0, x - K)$. Since $\sigma(B(T) - B(t))$ is distributed according to $\mathcal{N}(0, \sigma^2(T-t))$, we have that $\sigma(B(T) - B(t))$ is equal to $\sigma\sqrt{T-t} \cdot Y$, where $Y = (B(T) - B(t))/\sqrt{T-t} \sim \mathcal{N}(0,1)$. We calculate $\mathbb{E}\left[f(Z^{t,x}(T))\right]$:

$$\mathbb{E}\left[\max(0, Z^{t,x}(T) - K)\right] = \mathbb{E}\left[\max(0, e^{\ln Z^{t,x}(T)} - K)\right]$$
$$= \mathbb{E}\left[\max(0, e^{\ln x + (r-\sigma^2/2)(T-t) + \sigma\sqrt{T-t}\cdot Y} - K)\right].$$

Observe that the random variable inside the expectation is zero when Y is such that
$$\ln x + \left(r - \frac{1}{2}\sigma^2\right)(T-t) + \sigma\sqrt{T-t} \cdot Y < \ln K,$$
or, equivalently, when $Y < -d_2$, where
$$d_2 := \frac{\ln(x/K) + (r - \sigma^2/2)(T-t)}{\sigma\sqrt{T-t}}.$$

Below it will become clear why we define d_2 in this way. With this knowledge, we find

$$\mathbb{E}\left[\max(0, Z^{t,x}(T) - K)\right] = \int_{-d_2}^{\infty} \left(e^{\ln x + (r-\sigma^2/2)(T-t) + \sigma\sqrt{T-t}\cdot y} - K\right) \phi(y)\,dy$$
$$= xe^{r(T-t)} \int_{-d_2}^{\infty} e^{-\sigma^2(T-t)/2 + \sigma\sqrt{T-t}\cdot y} \phi(y)\,dy$$
$$- K \int_{-d_2}^{\infty} \phi(y)\,dy.$$

Since ϕ is the probability density of Y, we know that
$$\mathcal{P}(Y > -d_2) = \int_{-d_2}^{\infty} \phi(y)\,dy = \mathcal{P}(Y < d_2),$$
where the last equality follows from the symmetry of ϕ around zero. Hence, the second integral becomes
$$K \int_{-d_2}^{\infty} \phi(y)\,dy = K\Phi(d_2),$$
where we recall that Φ is the cumulative distribution function for a standard normal random variable. Considering the first integral, the change of variables $z = y - \sigma\sqrt{T-t}$ yields

$$\int_{-d_2}^{\infty} e^{-\sigma^2(T-t)/2+\sigma\sqrt{T-t}y}\phi(y)\,dy = \int_{-d_2}^{\infty} \frac{1}{\sqrt{2\pi}} e^{-(y-\sigma\sqrt{T-t})^2/2}\,dy$$
$$= \int_{-d_2-\sigma\sqrt{T-t}}^{\infty} \frac{1}{\sqrt{2\pi}} e^{-z^2/2}\,dz$$
$$= \Phi\left(d_2+\sigma\sqrt{T-t}\right).$$

After multiplying by the discounting factor $e^{-r(T-t)}$, we reach the Black & Scholes option pricing formula:

Theorem 4.6. *The price of a call option with strike K and exercise time T is*
$$P_t = S_t \Phi(d_1) - K e^{-r(T-t)} \Phi(d_2), \qquad (4.17)$$

where
$$d_1 = d_2 + \sigma\sqrt{T-t},$$
$$d_2 = \frac{\ln(S(t)/K) + (r - \sigma^2/2)(T-t)}{\sigma\sqrt{T-t}}.$$

Another representation of d_1 is
$$d_1 = \frac{\ln(S(t)/K) + (r + \sigma^2/2)(T-t)}{\sigma\sqrt{T-t}}.$$

This formula was derived by Fisher Black and Myron Scholes in their seminal work [9]. They evaluated the fair price of an option, and opened a whole new area of research together with Robert Merton [39]. The Black & Scholes option theory earned Merton and Scholes (F. Black died in 1995) the Nobel prize in economics in 1997. [5]

Notice that the price of the call option is independent of the drift α of the stock price dynamics. This is not surprising in view of the fact that α is nowhere present in the dynamics for $Z^{t,x}(s)$, nor in the Black & Scholes partial differential equation.

A call option is a financial contract whose value is a function of the volatility of the underlying stock, and in a popular language we may say that trading in options is the same as trading in volatility. Figure 4.1 exhibits the dependency of the call option price on volatility, where we have chosen $S(0) = 100$, strike $K = 100$ and $T = 50$ days to exercise. The yearly risk-free interest rate is 5%, which corresponds to $r = 0.000198$ on a daily time scale.[6] We varied the range of the volatilty from 0.001 up to 1, and we observe that the price increases from almost zero for very low volatility to 100 for extremely high volatilites. Using the Black & Scholes formula one can do analysis of the

---

[5] See www.nobel.se/economics/laureates/1997/index.html of the Nobel Institute for the prize award.
[6] Recall that we assume 252 trading days in a year.

4.3 Pricing and Hedging of Contingent Claims $X = f(S(T))$

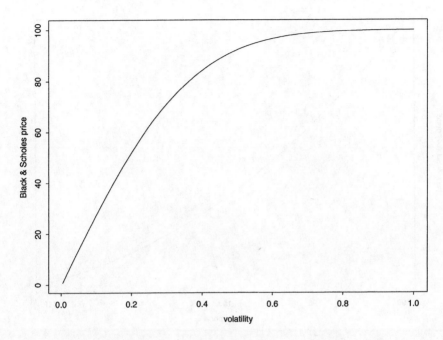

Fig. 4.1. The price of a call option as a function of volatility

option price asymptotics when the volatility goes to zero or to infinity (see Exercise 4.8). Considering Fig. 4.1 from a practical perspective, the volatility for stocks is "normally" between 15% to 30% yearly, which means a daily volatility between 0.01 and 0.02, approximately.

The option price is decreasing as a function of the strike K and increasing as a function of time T to exercise (check out Exercise 4.8). This is displayed in Figs. 4.2 and 4.3, where we assume $S(0) = 100$, a daily volatility $\sigma = 0.015$ and the risk-free rate equal to 5% yearly. In Fig. 4.2, time to exercise is chosen to be $T = 50$ days, while in Fig. 4.3 we kept the strike price fixed and equal to 100.

4.3.4 Hedging of Call Options

As we calculated the price of a call option, we can find its replicating portfolio. The number of shares in the hedge is $a^H(t) = \partial C(t, S(t))/\partial x$, and we thus need to calculate $\partial C(t,x)/\partial x$. By commuting expectation and differentiation we reach the following:

$$\begin{aligned} \frac{\partial C(t,x)}{\partial x} &= \frac{\partial}{\partial x} e^{r(T-t)} \mathbb{E}\left[f(Z^{t,x}(T))\right] \\ &= e^{-r(T-t)} \mathbb{E}\left[f'(Z^{t,x}(t))Z^{t,1}(T)\right]. \end{aligned}$$

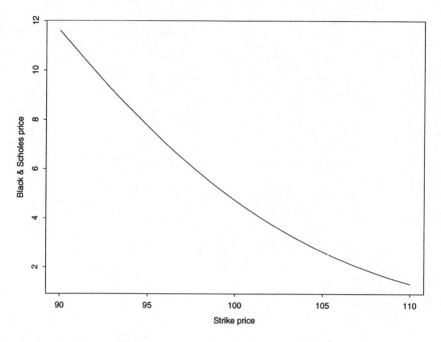

Fig. 4.2. The price of a call option as a function of the strike

The last equality follows from

$$\frac{\partial}{\partial x} Z^{t,x}(T) = \frac{\partial}{\partial x} x \cdot \exp\left(\left(r - \frac{1}{2}\sigma^2\right)(T-t) + \sigma(B(T) - B(t))\right)$$
$$= 1 \cdot \exp\left(\left(r - \frac{1}{2}\sigma^2\right)(T-t) + \sigma(B(T) - B(t))\right)$$
$$= Z^{t,1}(t).$$

Since we are replicating a call option, the function $f(x)$ is given by $f(x) = \max(0, x - K)$, and hence,

$$f'(x) = \begin{cases} 1, & x > K, \\ 0, & x < K. \end{cases}$$

Following the calculations for the price, we derive

$$\frac{\partial C(t,x)}{\partial x} = e^{-r(T-t)} \mathbb{E}\left[Z^{t,1}(T) 1_{\{Z^{t,x}(T) > K\}}\right]$$
$$= e^{-r(T-t)} \int_{-d_2}^{\infty} \frac{1}{\sqrt{2\pi}} e^{(r-\sigma^2/2)(T-t) + \sigma\sqrt{T-t}\,y - y^2/2}\, dy$$
$$= \int_{-\infty}^{d_1} \frac{1}{\sqrt{2\pi}} e^{-y^2/2}\, dy = \Phi(d_1).$$

4.3 Pricing and Hedging of Contingent Claims $X = f(S(T))$

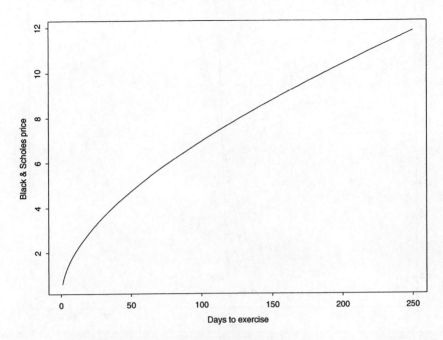

Fig. 4.3. The price of a call option as the function of exercise time

We have proved the following theorem.

Theorem 4.7. *The number of shares of the underlying stock in the hedging portfolio of a call option with strike price K at exercise time T is given by $a^H(t) = \Phi(d_1)$, where d_1 is defined in Thm. 4.6.*

Recall that Φ is the cumulative probability function of a standard normal distribution, and therefore $0 \leq a^H(t) \leq 1$. The hedging position in the stock should always be less than one, but greater than zero, and is therefore often referred to as the *hedge ratio*. In Fig. 4.4 we plot the hedge ratio $a^H(t)$ for a call option as a function of the price of the underlying stock for several different times in order to exhibit how it changes with the time up to exercise. We assume that the option has a strike price $K = 100$ and exercise time is $T = 100$ days. The volatility is put equal to 0.015, while the interest rate is set to $r = 0$. From the figure we see that just before exercise the hedging portfolio consists of approximately one stock if stock price is above strike, and no stocks otherwise. This coincides with our intuition saying that there is no reason to hedge if the option is not likely to be exercised.

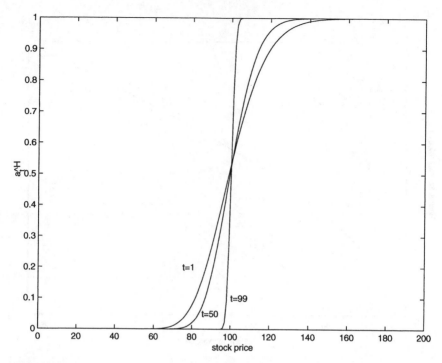

Fig. 4.4. Number of shares in the replicating portfolio for a call option as a function of the underlying stock price

4.3.5 Hedging of General Options

In the derivation of the hedging strategy for a call option we differentiated the payoff function $f(x)$. This approach does not always work and must be used with care. Consider, for example, a digital option which pays €1 if the value of a stock is bigger than a strike price K at exercise time T, and zero otherwise. Such an option contract has the payoff function $f(x) = 1_{x>K}$, a function which has derivative equal to zero except at $x = K$, where the derivative is undefined. Note the similarity to a call option, which also has a point where the derivative is undefined. One may believe that since the digital payoff function has zero derivative except at one point, the hedge will consist of zero shares of the underlying stock. This cannot be true, of course. We shall show an alternative route to find the delta of an option based on the so-called *density approach*. In words, this approach moves the dependency of the initial state of the stock x from the payoff function to the density function when calculating the expectation in (4.16). In this way the differentiation with respect to x of the option price will not involve the payoff function itself. We note that this method is a special case of the much more general *Malliavin*

4.3 Pricing and Hedging of Contingent Claims $X = f(S(T))$

approach that we encounter in Sect. 4.5 when discussing hedging of general claims. Let us now go into details about the density approach.

Recall that

$$Z^{t,x}(T) = x \exp\left(\left(r - \frac{1}{2}\sigma^2\right)(T-t) + \sigma(B(T) - B(t))\right),$$

which can be written as $Z^{t,x}(T) = \exp(Y)$ for a random variable $Y = \ln x + (r - 0.5\sigma^2)(T-t) + \sigma(B(T) - B(t))$ distributed as

$$Y \sim \mathcal{N}\left(\ln x + \left(r - \frac{1}{2}\sigma^2\right)(T-t), \sigma^2(T-t)\right).$$

The density of Y is denoted by $p_Y(y; x)$ and is explicitly written as

$$p_Y(y; x) = \frac{1}{\sqrt{2\pi\sigma^2(T-t)}} \exp\left(-\frac{(y - \ln x - (r - \sigma^2/2)(T-t))^2}{2\sigma^2(T-t)}\right).$$

Note that we use the notation $p_Y(y; x)$ rather than the standard $p_Y(y)$ to emphasize the dependence on x.

From (4.16) we can write the price of the option as

$$C(t,x) = e^{-r(T-t)} \mathbb{E}\left[f(Z^{t,x}(T))\right] = e^{-r(T-t)} \int_{\mathbb{R}} f(e^y) p_Y(y; x) \, dy.$$

We calculate the delta of the option:

$$\frac{\partial C(t,x)}{\partial x} = \frac{\partial}{\partial x} e^{-r(T-t)} \int_{\mathbb{R}} f(e^y) p_Y(y; x) \, dy$$

$$= e^{-r(T-t)} \int_{\mathbb{R}} f(e^y) \frac{\partial p_Y(y; x)}{\partial x} \, dy$$

$$= e^{-r(T-t)} \int_{\mathbb{R}} f(e^y) p_Y(y; x) \frac{y - \ln x - (r - \sigma^2/2)(T-t)}{x\sigma^2(T-t)} \, dy$$

$$= e^{-r(T-t)} \mathbb{E}\left[f(Z^{t,x}(T)) \frac{\ln Z^{t,x}(T) - \ln x - (r - \sigma^2/2)(T-t)}{x\sigma^2(T-t)}\right].$$

Here we used $Y = \ln Z^{t,x}(T)$ and

$$\frac{\partial p_Y(y; x)}{\partial x} = p_Y(y; x) \cdot \frac{y - \ln x - (r - \sigma^2/2)(T-t)}{x\sigma^2(T-t)}.$$

We conclude the density approach to hedging in the following theorem.

Theorem 4.8. *The delta (or hedge ratio) of a contingent claim with payoff $X = f(S(T))$ is given by $a^H(t) = \partial C(t, S(t))/\partial x$, where*

$$\frac{\partial C(t,x)}{\partial x} = e^{-r(T-t)} \mathbb{E}\left[g(t, x, Z^{t,x}(T))\right],$$

for a "payoff" function g defined as

$$g(t,x,z) = f(z)\frac{\ln z - \ln x - (r - \sigma^2/2)(T-t)}{x\sigma^2(T-t)}.$$

We observe that the delta of the claim is simply the price of an option with payoff function g. Note that g depends on the option's payoff function f, but not the derivative of this. This is the advantage of the density method. Note that the number of shares in the hedge of a digital option will be different from zero (see Exercise 4.12).

4.3.6 Implied Volatility

We will now spend a few lines discussing *implied volatility*. We discussed in Sect. 2.2 how to estimate the volatility σ from historical stock prices. The volatility σ is the only parameter which is unknown to us when pricing a call option contract. The risk-free rate of return r is taken from the return of the Treasury bills, while K and T are contractual parameters. Today's stock price $S(0)$ is obviously known to the investors. The critical parameter for deriving a fair call option price is therefore σ. By inverting the Black & Scholes formula one can find out what volatility the investors use when trading in call options. Let us go into this in more detail.

Suppose we know that a call option with strike K and time to exercise T is traded for a price p in the market. At the same time, we read off from the stock exchange monitor that the underlying stock is traded for price s. If the risk-free rate of return is r, we know from the Black & Scholes formula in Thm. 4.6 that

$$p = s\Phi(d_1) - Ke^{-rT}\Phi(d_2),$$

where $d_1 = d_2 + \sigma\sqrt{T}$ and

$$d_2 = \frac{\ln(s/K) + (r - \sigma^2/2)T}{\sigma\sqrt{T}}.$$

Since the only unknown here is σ, we can solve for this and find the volatility used by the market. Ideally, this should coincide (at least approximately) with the historical volatility, but this is rarely the case. Since this volatility is derived *from* actual option prices, we call it the *implied volatility*. It is often used for getting a picture of the current level of volatility, and as the basis for pricing new options on stocks that are not so frequently traded but where more liquid options can be used as proxies.

Unfortunately, it is not possible to derive a closed-form solution for the implied volatility, so we need to resort to numerical estimation techniques. One simple way to do this is to implement the Black & Scholes formula in Excel and use the "Solver" optionality. Another, far more cumbersome way is by trial and error.

Most often several options are traded on the same underlying stock. For instance, the market can trade in options on the same underlying stock but with different strike prices for the same exercise time T. Consider, for example, n options traded with strikes K_i, $i = 1, 2, ..., n$. Denote the respective market quoted prices by p_i, $i = 1, 2, ..., n$. Each option will now yield an implied volatility, σ_i, $i = 1, 2, ..., n$. In an ideal market, we would have $\sigma_1 = \sigma_2 = ... = \sigma_n$, but this is usually not the case. If the Black & Scholes hypotheses about a complete arbitrage-free market with no frictions hold, these implied volatilities would be equal and coincide with the historically estimated one. What we observe, however, is a *volatility smile*. In Fig. 4.5 we display the implied volatilities from call options on Microsoft quoted on NASDAQ June 9, 2003. We calculated the implied volatilities from calls with exercise date October 17, 2003, which means 94 trading days to exercise. The calls had strike prices ranging from \$22.5 to \$37.5, and the underlying Microsoft stock was traded for \$23.75. On this day the risk-free interest rate on US Treasury bills with 6 months to maturity was 0.92%. We used Solver in Excel to produce the implied volatilities that we see in Fig. 4.5. The volatilities are measured on a daily scale, and for comparison we estimate the historical volatility to be $\hat{\sigma} = 0.02666$ (which corresponds to 42.3% yearly volatility). The volatility smile is maybe not among the sweetest, but the idea behind the name is hopefully clear. Note that the implied volatilities range from 30.2% up to 42.0% yearly, all being much lower than the historical volatility for this specific example. The volatility smile is not static, so when time progresses towards the exercise date, the smile will also change shape.

4.4 The Girsanov Theorem and Equivalent Martingale Measures

We claimed in Sect. 4.1 that the arbitrage-free price of a derivative X is the discounted expected payoff, where the expectation is taken under the risk-neutral probability. Formula (4.16) is not far away from being a risk-neutral present value of X, and it is tempting to try to establish such a representation for the price in the Black & Scholes market. This is the objective of the current subsection, where the Girsanov theorem will be the key to constructing risk-neutral probabilities.

To establish what our aim is, define the stochastic process $dW(t) := dB(t) + \frac{\alpha - r}{\sigma} dt$ and substitute $dB(t)$ with $dW(t)$ in the dynamics of $S(t)$. We find
$$dS(t) = \alpha S(t)\, dt + \sigma S(t) \left(dW(t) - \frac{\alpha - r}{\sigma}\, dt \right),$$
or,
$$dS(t) = rS(t)\, dt + \sigma S(t)\, dW(t),$$
which is a dynamics very similar to that of $Z(t)$. However, $W(t)$ is not a Brownian motion. For instance,

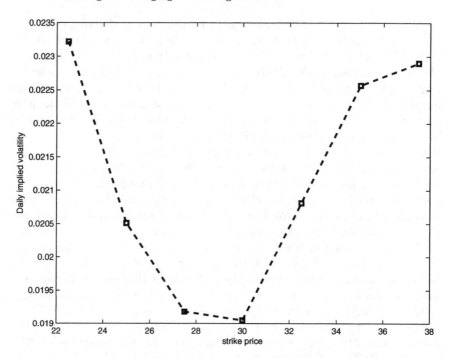

Fig. 4.5. Implied volatility as a function of the strike price calculated from call options on Microsoft quoted on NASDAQ June 9, 2003 (depicted as squares)

$$\mathbb{E}\left[W(t)\right] = \frac{\alpha - r}{\sigma} t,$$

which is not zero (or constant with respect to time). When we introduced the Brownian motion $B(t)$, we assigned probabilistic properties like independent and normally distributed stationary increments. These properties refer to the probability $\mathcal{P}$ on Ω. So, when claiming that $W(t)$ is not a Brownian motion, we really mean that it is not a Brownian motion *with respect to the probability* $\mathcal{P}$. We may ask, however, if there exists *some other* probability on Ω where $W(t)$ *is* a Brownian motion. The answer is positive, and this probability is defined via the Girsanov theorem. It will be the analogue to our risk-neutral probability in the one-period market. In the Black & Scholes market the name *equivalent martingale measure* is given to it.

Let us first define what we mean by an equivalent martingale measure on Ω.

Definition 4.9. *A probability $\mathcal{Q}$ is called an* equivalent martingale measure *if there exists a random variable $Y > 0$ such that $\mathcal{Q}(A) = \mathbb{E}\left[1_A Y\right]$ for all events A and $e^{-rt} S(t)$ is a martingale with respect to $\mathcal{Q}$.*

Let us investigate the reason for the rather cumbersome name associated with such a probability $\mathcal{Q}$. First, *equivalent* means that if $\mathcal{P}(A) > 0$ for an

4.4 The Girsanov Theorem and Equivalent Martingale Measures

event $A \subset \Omega$, then $\mathcal{Q}(A) > 0$, and vice versa. Thus, what $\mathcal{P}$ predicts with positive probability, also $\mathcal{Q}$ predicts with positive probability. The opposite also holds. The name *martingale* obviously comes from the martingale property of the discounted stock price. Notice that both the stock and the bond are martingales after discounting, being the two investment opportunities for hedging. The discounted bond price is simply the constant 1. Finally, *measure* is a generalization of the notion of probability, and is used even though $\mathcal{Q}$ is a probability. Frequently, an equivalent martingale measure $\mathcal{Q}$ is called a *risk-neutral probability* since the expected rate of return from an investment in the stock is r with respect to $\mathcal{Q}$.

We are now ready to state the Girsanov theorem.[7]

Theorem 4.10. *For $\lambda \in \mathbb{R}$, define the probability*

$$\mathcal{Q}(A) := \mathbb{E}\left[1_A M(T)\right], \quad A \subset \Omega, \tag{4.18}$$

where

$$M(T) = \exp\left(-\lambda B(T) - \frac{1}{2}\lambda^2 T\right).$$

Then $W(t) := B(t) + \lambda t$ is a Brownian motion for $0 \le t \le T$ with respect to $\mathcal{Q}$.

Choosing $\lambda = \frac{\alpha - r}{\sigma}$, we see that

$$W(t) = B(t) + \frac{\alpha - r}{\sigma} t,$$

becomes a Brownian motion under the probability $\mathcal{Q}$ defined in Girsanov's theorem 4.10. We shall stick to this choice of λ in the rest of this chapter. Using Itô's formula it is straightforward to show that the discounted price $S^{\mathrm{d}}(t) := \mathrm{e}^{-rt} S(t)$ has the dynamics

$$\mathrm{d}S^{\mathrm{d}}(t) = \sigma S^{\mathrm{d}}(t)\,\mathrm{d}W(t),$$

or, equivalently,

$$S^{\mathrm{d}}(t) = S(0) + \int_0^t \sigma S^{\mathrm{d}}(u)\,\mathrm{d}W(u),$$

which demonstrates that $S^{\mathrm{d}}(t)$ is a martingale with respect to $\mathcal{Q}$. Hence, letting $Y = M(T)$ in Def. 4.9 above, we find that $\mathcal{Q}$ in Thm. 4.10 is an equivalent martingale measure. Furthermore, since this probability is the only one that turns $W(t)$ into a Brownian motion, it is the unique equivalent martingale measure in the Black & Scholes market. Readers may convince themselves that there is only one way for the discounted price process to be a martingale, and that is when $W(t)$ is a Brownian motion.

---

[7] In fact, this is a special case of Girsanov's theorem. See, for instance, [33, 41] for a general version.

We now have two "parallel probability universes". The original probability space is $(\Omega, \mathcal{P})$, where our stock price model is defined and $B(t)$ is a Brownian motion. In parallel we have the risk-neutral world defined by the probability space $(\Omega, \mathcal{Q})$, where $W(t)$ is a Brownian motion. Notice that $B(t) = W(t) - \frac{\alpha-r}{\sigma}t$ is not a Brownian motion in this "universe". The purpose of the former probability space is to model stocks, while the latter is tailor-made for pricing options.

If we denote by $S^{t,x}(T)$ the stock price at time T when it starts at time t with value x, we see that $S^{t,x}(T)$ has the same distribution with respect to $\mathcal{Q}$ as $Z^{t,x}(T)$ has with respect to $\mathcal{P}$ since

$$S^{t,x}(T) = x\exp\left(\left(r - \frac{1}{2}\sigma^2\right)T + \sigma W(T)\right),$$

$$Z^{t,x}(T) = x\exp\left(\left(r - \frac{1}{2}\sigma^2\right)T + \sigma B(T)\right).$$

Hence, we can write (4.16) as in the theorem below.

Theorem 4.11. *The price of a contingent claim with payoff $X = f(S(T))$ is given by $P_t = C(t, S(t))$, where*

$$C(t, x) = e^{-r(T-t)}\mathbb{E}_\mathcal{Q}\left[f(S^{t,x}(T))\right]. \tag{4.19}$$

The expectation $\mathbb{E}_\mathcal{Q}$ is with respect to the probability $\mathcal{Q}$ defined in Thm. 4.10 with $\lambda = (\alpha - r)/\sigma$.

From the density approach to deriving the delta of a claim, we find the following result.

Theorem 4.12. *The delta (or hedge ratio) of a contingent claim with payoff $X = f(S(T))$ is given by $a^H(t) = \partial C(t, S(t))/\partial x$ where*

$$\frac{\partial C(t,x)}{\partial x} = e^{-r(T-t)}\mathbb{E}_\mathcal{Q}\left[g(t, x, S^{t,x}(T))\right],$$

for a "payoff" function g defined as

$$g(t, x, s) = f(s)\frac{\ln s - \ln x - (r - \sigma^2/2)(T-t)}{x\sigma^2(T-t)}.$$

The expectation $\mathbb{E}_\mathcal{Q}$ is with respect to the probability $\mathcal{Q}$ defined in Thm. 4.10 with $\lambda = (\alpha - r)/\sigma$.

A final remark on the stochastic variable $M(T)$ is necessary in order to complete this section. The notation indicates that $M(t)$ is a martingale, and indeed this is true (see Exercise 4.15). This property of $M(t)$ is crucial in the proof of Girsanov's theorem, which we will not present here. The interested reader can take a look at [33, 41] for the complete arguments leading to this key result in finance. To build up some knowledge on why $W(t)$ is a Brownian motion with respect to $\mathcal{Q}$, the reader should try to solve Exercise 4.16.

4.5 Pricing and Hedging of General Contingent Claims

In the introduction to this chapter we mentioned two examples of contingent claims with payoffs that cannot be expressed solely as a function of the terminal stock price. Asian options and barrier options are so-called *path dependent* options, and their arbitrage-free price can be calculated using the *martingale approach*. The goal of this section is to derive price dynamics for general contingent claims, and look into the problem of hedging. Unfortunately, to find the replicating strategy we must introduce some heavy machinery from stochastic analysis. We shall refrain from doing so, and be content with some indications in this direction. Let us start the presentation of the martingale approach to option pricing.

First, we recall the equivalent martingale measure $\mathcal{Q}$ introduced in the previous section by the Girsanov theorem, and assume that the payoff from the contingent claim X satisfies the moment condition $\mathbb{E}_\mathcal{Q}\left[X^2\right] < \infty$. Our market is assumed to be complete, so there exists a hedging strategy (a^H, b^H) replicating our claim. From the assumption of an arbitrage-free market, we have (like in Sect. 4.3) that $P(t) = H(t)$, where $H(t)$ is the value of the hedge and $P(t)$ is the price of the contingent claim at time t.

We claim that the discounted value of the hedging portfolio $H^\mathrm{d}(t) := \mathrm{e}^{-rt} H(t)$ is a martingale with respect to $\mathcal{Q}$. Before we demonstrate that this is indeed true, let us see what are the consequences of such a result. In fact, the martingale property of the discounted hedging portfolio is the key to pricing and hedging of claims. This is shown by the following little argument. Since $H(t)$ is the hedging portfolio, we have $H(T) = X$. This implies of course that $H^\mathrm{d}(T) = \mathrm{e}^{-rT} H(T) = \mathrm{e}^{-rT} X$. But from the martingale property of $H^\mathrm{d}(t)$ it follows that

$$H^\mathrm{d}(t) = \mathbb{E}_\mathcal{Q}\left[H^\mathrm{d}(T) \mid \mathcal{F}_t\right] = \mathbb{E}_\mathcal{Q}\left[\mathrm{e}^{-rT} X \mid \mathcal{F}_t\right].$$

Note that the conditional expectation introduced in Sect. 3.4 was with respect to $\mathcal{P}$. The definition is exactly the same when we work with the probability $\mathcal{Q}$ except for obvious modifications. Using the argument of no arbitrage, we know that $H^\mathrm{d}(t) = \mathrm{e}^{-rt} P(t)$. Hence,

$$\mathrm{e}^{-rt} P(t) = \mathrm{e}^{-rT} \mathbb{E}_\mathcal{Q}\left[X \mid \mathcal{F}_t\right].$$

After multiplying both sides by $\exp(rt)$, we can conclude with the theorem below.

Theorem 4.13. *The price of a contingent claim X at time t is*

$$P(t) = \mathrm{e}^{-r(T-t)} \mathbb{E}_\mathcal{Q}\left[X \mid \mathcal{F}_t\right]. \tag{4.20}$$

The price at $t = 0$ is

$$P(0) = \mathrm{e}^{-rT} \mathbb{E}_\mathcal{Q}\left[X\right]. \tag{4.21}$$

The last conclusion in the theorem follows from the properties of conditional expectation.

The whole argument leading to Thm. 4.13 rests on the martingale property of $H^d(t)$. By definition

$$H^d(t) = a^H(t)(e^{-rt}S(t)) + b^H(t)(e^{-rt}R(t)) = a^H(t)S^d(t) + b^H(t) \times 1.$$

We recall from the previous section that $dS^d(t) = \sigma S^d(t)\,dt$. It is also clear that $d(1) = 0$ since the constant 1 does not change with time. In Exercise 4.17 we show that under the self-financing hypothesis we have

$$dH^d(t) = a^H(t)dS^d(t) = a^H(t)\sigma S^d(t)\,dW(t),$$

which proves the martingale property. Hence, the conclusions of Thm. 4.13 rest on firm ground.

An important observation from Thm. 4.13 is that the discounted price process of the claim $P^d(t) := e^{-rt}P(t)$ is a martingale with respect to $\mathcal{Q}$. Hence, we see that in our market *all* the tradeable assets (bond, stock and claim) are martingales after discounting with respect to the equivalent martingale measure $\mathcal{Q}$. This in turn implies that there are no arbitrage opportunites in the market. Assume conversely that (a, b, c) is an investment strategy such that $V(0) \leq 0$ and $V(T) \geq 0$ with $\mathbb{E}\left[V(T)\right] > 0$. From the martingale property of the discounted price processes it follows that $V^d(t) := e^{-rt}V(t)$ is a martingale with respect to $\mathcal{Q}$ (this can be proved using the same arguments as in Exercise 4.17). Therefore,

$$0 \geq V(0) = e^{-rT}\mathbb{E}_{\mathcal{Q}}\left[V(T)\right].$$

Hence, $\mathbb{E}_{\mathcal{Q}}\left[V(T)\right] \leq 0$ which from the non-negativity of $V(T)$ gives $\mathcal{Q}(V(T) > 0) = 0$. But $\mathcal{Q}$ and $\mathcal{P}$ are equivalent, implying $\mathcal{P}(V(T) > 0) = 0$ because they agree on all impossible events. Hence, $\mathcal{P}(V(T) \leq 0) = 1$, which contradicts $\mathbb{E}\left[V(T)\right] > 0$. What we have proved is that if there exists an equivalent probability $\mathcal{Q}$ such that *all* tradeable assets in the market are martingales after discounting, the market is free of arbitrage. Suprisingly enough, the opposite implication is also valid: if the market is arbitrage-free, there exists an equivalent martingale measure. This will be discussed further in Sect. 4.8.

The practically oriented reader may have asked about the applicability of the result in Thm. 4.13. Admittedly, the conditional expectation is not exactly easy to calculate, and in some sense we are left with a pricing dynamics which looks mathematically nice, but is practically worthless. This is not entirely true. The expression paves the way for a theoretical study of the price dynamics. But even more interestingly, it is the starting point for numerical methods (see [27]). Also, in many circumstances, one can derive a closed-form expression for the conditional expectation. Lastly, the pricing dynamics is independent of the hedging strategy, even though the whole derivation made use of this.

4.5 Pricing and Hedging of General Contingent Claims

We now move on to discuss the replication of general claims. Since $P^{\mathrm{d}}(t) = H^{\mathrm{d}}(t)$, where $\mathrm{d}H^{\mathrm{d}}(t) = \sigma a^H(t) S^{\mathrm{d}}(t)\, \mathrm{d}W(t)$, we have from Thm. 4.13

$$\mathrm{e}^{-rT}\mathbb{E}_{\mathcal{Q}}\left[X \mid \mathcal{F}_t\right] = P(0) + \int_0^t \sigma a^H(u) S^{\mathrm{d}}(u)\, \mathrm{d}W(u).$$

This is an integral equation for the process $a^H(t)$, where we have a known stochastic process on the left-hand side, and an Itô integral of the unkown on the right-hand side. But how can one solve the hedging strategy from such a highly unorthodox integral equation? Let us for a moment suppose that we do not have an Itô integral on the right-hand side, but a standard integral over time:

$$\mathrm{e}^{-rT}\mathbb{E}_{\mathcal{Q}}\left[X \mid \mathcal{F}_t\right] = P(0) + \int_0^t \sigma a^H(u) S^{\mathrm{d}}(u)\, \mathrm{d}u.$$

If we differentiate both sides with respect to time, we obtain the expression

$$\sigma a^H(t) S^{\mathrm{d}}(t) = \frac{\mathrm{d}}{\mathrm{d}t}\left(\mathrm{e}^{-rT}\mathbb{E}_{\mathcal{Q}}\left[X \mid \mathcal{F}_t\right]\right),$$

or,

$$a^H(t) = (\sigma S^{\mathrm{d}}(t))^{-1} \mathrm{e}^{-rT} \frac{\mathrm{d}}{\mathrm{d}t} \mathbb{E}_{\mathcal{Q}}\left[X \mid \mathcal{F}_t\right].$$

So, in the classical situation, we simply differentiate and end up with an expression for $a^H(t)$ after reorganizing. In our case we have an Itô integral instead of a standard time integral, and this differentiation will not be allowed, of course. However, one may ask if there exists a kind of differentiation which works as ordinary differentiation for time integrals, and the answer is affirmative. We have

$$a^H(t) = (\sigma S^{\mathrm{d}}(t))^{-1} \mathrm{e}^{-rT} \mathbb{E}_{\mathcal{Q}}\left[\mathrm{D}_t X \mid \mathcal{F}_t\right], \qquad (4.22)$$

where D_t is the so-called *Malliavin derivative*. Notice that we have put the differentiation operator D_t *inside* the expectation, which may seem somewhat surprising. The Malliavin derivative is based on advanced stochastic analysis and we are not going to explain in any more detail what the right-hand side of (4.22) means. Those readers who are interested in this exciting theory and its applications to finance are referred to the lecture notes [42]. For many claims it is possible to calculate the right-hand side of (4.22) and obtain an expression that can be used as the basis for practical hedging strategies. For advanced financial applications of the Malliavin derivative to Asian options and other exotic claims we refer to [27].

4.5.1 An Example: a Chooser Option

Let us consider a *chooser option* on a stock, a derivative that gives holders the right to *choose* at a contracted time $t < T$ if they want a call or put option.

If the exercise time of the chooser option is T with strike K, the holder will either have a call or a put at time T, depending on the choice made at the earlier time t. Of course, a rational investor will choose the option which is most valuable at time t. We are now going to derive the arbitrage-free price at time 0 for this option contract. As we will see in just a moment, the payoff function is not representable as a function of the terminal stock price, and the theory for general claims must be applied to find a price.

Let $P^c(t)$ and $P^p(t)$ be the price at time t for a call and put option respectively, with strike K and exercise at T. The holder of the chooser option will take the call option at time t if $P^c(t) \geq P^p(t)$, and the put option otherwise. The payoff function at time T is

$$X = \max\left(0, S(T) - K\right) 1_{\{P^c(t) \geq P^p(t)\}} + \max\left(0, K - S(T)\right) 1_{\{P^c(t) < P^p(t)\}}.$$

From the theory we find the arbitrage-free price $P(0)$ at the time of entering the chooser option to be

$$P(0) = e^{-rT} \mathbb{E}_{\mathcal{Q}}[X]. \tag{4.23}$$

Our task now is to calculate this price.

Adding and subtracting $\max(0, S(T) - K) 1_{\{P^c(t) < P^p(t)\}}$ in X, yields

$$X = \max\left(0, S(T) - K\right) + (K - S(T)) 1_{\{P^c(t) < P^p(t)\}},$$

since $\max(0, K-x) - \max(0, x-K) = K-x$ and $1_{\{P^c(t) \geq P^p(t)\}} + 1_{\{P^c(t) < P^p(t)\}} = 1$. Hence,

$$P(0) = e^{-rT} \mathbb{E}_{\mathcal{Q}}\left[\max(0, S(T) - K) + (K - S(T)) 1_{\{P^c(t) < P^p(t)\}}\right]$$
$$= e^{-rT} \mathbb{E}_{\mathcal{Q}}\left[\max(0, S(T) - K)\right] + e^{-rT} \mathbb{E}_{\mathcal{Q}}\left[(K - S(T)) 1_{\{P^c(t) < P^p(t)\}}\right].$$

We recognize the first term as the price of a call option with strike K at the exercise time T. Therefore,

$$P(0) = P^c(0) + e^{-rT} \mathbb{E}_{\mathcal{Q}}\left[(K - S(T)) 1_{\{P^c(t) < P^p(t)\}}\right].$$

We calculate the second term on the right-hand side.

Recalling the law of double expectation, we find after conditioning on $\mathcal{F}_t$,

$$e^{-rT} \mathbb{E}_{\mathcal{Q}}[(K - S(T)) 1_{\{P^c(t) < P^p(t)\}}]$$
$$= \mathbb{E}_{\mathcal{Q}}\left[e^{-rT}(K - S(T)) 1_{\{P^c(t) < P^p(t)\}}\right]$$
$$= \mathbb{E}_{\mathcal{Q}}\left[\mathbb{E}_{\mathcal{Q}}\left[e^{-rT}(K - S(T)) 1_{\{P^c(t) < P^p(t)\}} \mid \mathcal{F}_t\right]\right].$$

The Black & Scholes formula tells us that $P^c(t)$ is $\mathcal{F}_t$-adapted since it is a function of the stock price at time t. Further, the put–call parity that the reader is asked to prove in Exercise 4.10 gives

$$P^p(t) = P^c(t) - S(t) + Ke^{-r(T-t)}, \tag{4.24}$$

which therefore implies the $\mathcal{F}_t$-adaptedness of $P^{\mathrm{p}}(t)$. In conclusion, the random variable $1_{\{P^{\mathrm{c}}(t) < P^{\mathrm{p}}(t)\}}$ is only dependent on the stock price at time t, and therefore $\mathcal{F}_t$-adapted. From the properties of conditional expectation in Sect. 3.4, we can move $1_{\{P^{\mathrm{k}}(t) < P^{\mathrm{s}}(t)\}}$ outside the conditioning

$$\mathbb{E}_{\mathcal{Q}}[\mathbb{E}_{\mathcal{Q}}[\mathrm{e}^{-rT}(K - S(T))1_{\{P^{\mathrm{c}}(t) < P^{\mathrm{p}}(t)\}} \mid \mathcal{F}_t]]$$
$$= \mathbb{E}_{\mathcal{Q}}\left[1_{\{P^{\mathrm{c}}(t) < P^{\mathrm{p}}(t)\}} \mathbb{E}_{\mathcal{Q}}\left[\mathrm{e}^{-rT}(K - S(T)) \mid \mathcal{F}_t\right]\right]$$
$$= \mathbb{E}_{\mathcal{Q}}\left[1_{\{P^{\mathrm{c}}(t) < P^{\mathrm{p}}(t)\}}\left(\mathrm{e}^{-rT}K - \mathbb{E}_{\mathcal{Q}}\left[\mathrm{e}^{-rT}S(T) \mid \mathcal{F}_t\right]\right)\right]$$
$$= \mathbb{E}_{\mathcal{Q}}\left[1_{\{P^{\mathrm{c}}(t) < P^{\mathrm{p}}(t)\}}\left(\mathrm{e}^{-rT}K - \mathrm{e}^{-rt}S(t)\right)\right].$$

In the last step we used the martingale property with respect to $\mathcal{Q}$ of the discounted stock price $\mathrm{e}^{-rT}S(T)$. Simple factorization leads to

$$\mathbb{E}_{\mathcal{Q}}[\mathrm{e}^{-rT}(K - S(T))1_{\{P^{\mathrm{c}}(t) < P^{\mathrm{p}}(t)\}}]$$
$$= \mathrm{e}^{-rt}\mathbb{E}_{\mathcal{Q}}\left[(K\mathrm{e}^{-r(T-t)} - S(t))1_{\{P^{\mathrm{c}}(t) < P^{\mathrm{p}}(t)\}}\right].$$

We investigate when $P^{\mathrm{c}}(t) < P^{\mathrm{p}}(t)$. If $\omega \in \Omega$ is such that $P^{\mathrm{c}}(t, \omega) < P^{\mathrm{p}}(t, \omega)$, the put–call parity (4.24) implies that $S(t, \omega) < K\mathrm{e}^{-r(T-t)}$. The opposite implication also holds, and the two events $\{\omega \in \Omega \mid P^{\mathrm{c}}(t, \omega) < P^{\mathrm{p}}(t, \omega)\}$ and $\{\omega \in \Omega \mid S(t, \omega) < K\mathrm{e}^{-r(T-t)}\}$ are identical. We get

$$\mathbb{E}_{\mathcal{Q}}[\mathrm{e}^{-rT}(K - S(T))1_{\{P^{\mathrm{c}}(t) < P^{\mathrm{p}}(t)\}}]$$
$$= \mathrm{e}^{-rt}\mathbb{E}_{\mathcal{Q}}\left[\left(K\mathrm{e}^{-r(T-t)} - S(t)\right)1_{\{S(t) < K\mathrm{e}^{-r(T-t)}\}}\right]$$
$$= \mathrm{e}^{-rt}\mathbb{E}_{\mathcal{Q}}\left[\max\left(0, K\mathrm{e}^{-r(T-t)} - S(t)\right)\right].$$

We recognize the last term as the price of a put option with exercise time t and strike price $K\mathrm{e}^{-r(T-t)}$.

In conclusion, the chooser option has an arbitrage-free price which is a sum of a call option with exercise K at time T and a put option with exercise time t and strike $K\mathrm{e}^{-r(T-t)}$.

$$P(0) = P^{\mathrm{c}}(0) + \mathrm{e}^{-rt}\mathbb{E}_{\mathcal{Q}}\left[\max\left(0, K\mathrm{e}^{-r(T-t)} - S(t)\right)\right]. \quad (4.25)$$

It is left to the reader to derive a Black & Scholes-type formula for the chooser option contract, and also to show that the price can be represented as the sum of a put option with exercise time T and strike K and a call option with exercise time t and strike $K\mathrm{e}^{-r(T-t)}$ (see Exercise 4.18).

4.6 The Markov Property and Pricing of General Contingent Claims

Since claims with payoff $X = f(S(T))$ are special cases of general contingent claims, the general pricing theory from the previous section must coincide with the prices derived in Thm. 4.11, that is,

$$e^{-r(T-t)} \mathbb{E}_{\mathcal{Q}}\left[f(S(T)) \mid \mathcal{F}_t\right] = C(t, S(t)),$$

where
$$C(t, x) = e^{-r(T-t)} \mathbb{E}_{\mathcal{Q}}\left[f(S^{t,x}(T))\right].$$

The key to establishing this relation is the *Markov property* of geometric Brownian motion.

An adapted stochastic process $X(t)$ is called a Markov process if the future behaviour *only* depends on the current state of the process, and not on its past. Translated into a financial context, the Markov property of the stock price dynamics means that *all* information is contained in the current stock price only, and knowledge about how the stock has reached its current state will not provide us with any additional information when predicting the future states of stock prices. We define mathematically the Markov property in a way suitable for our purposes.

Definition 4.14. *An adapted stochastic process $X(t)$ is called a* Markov process *if for every $s > t \geq 0$*

$$\mathbb{E}\left[f(X(s)) \mid \mathcal{F}_t\right] = g(t, s, X^{0,x}(t)), \tag{4.26}$$

for all real-valued functions f such that $\mathbb{E}\left[|f(X(t))|\right] < \infty$, where

$$g(t, s, y) = \mathbb{E}\left[f(X^{t,y}(s))\right]. \tag{4.27}$$

When $g(t, s, y) = g(0, s - t, y)$, we say that $X(t)$ is a time-homogeneous Markov process.

Recall that $\mathcal{F}_t$ consists of all information that Brownian motion is generating up to time t. Hence, it contains the information generated by the process $X(s)$ up to time t since this is adapted. The definition of the Markov property then states that conditioning on all the possible paths of Brownian motion (or $X(s)$) up to time t is the same as conditioning on that the process starts in its current state at time t, namely $X^{0,x}(t)$. The reader should be aware that martingale and Markov properties are *not* the same.

Geometric Brownian motion $S(t)$ is a time-homogeneous Markov process. We shall not go into the details to show this, but simply note that for $s > t$ we have

$$\begin{aligned} S(s) &= S(0) \exp\left(\left(\alpha - \frac{1}{2}\sigma^2\right) s + \sigma B(s)\right) \\ &= S(0) \exp\left(\left(\alpha - \frac{1}{2}\sigma^2\right) t + \sigma B(t)\right) \\ &\quad \times \exp\left(\left(\alpha - \frac{1}{2}\sigma^2\right) (s - t) + \sigma(B(s) - B(t))\right) \\ &= S(t) \exp\left(\left(\alpha - \frac{1}{2}\sigma^2\right) (s - t) + \sigma(B(s) - B(t))\right). \end{aligned}$$

Note that the Brownian increment $B(s) - B(t)$ is independent of $S(t)$, and hence, $S(s)$ is the product of the current state $S(t)$ and an independent random variable. Therefore we see that all information about geometric Brownian motion at time t is contained in the current state, and conditioning on $\mathcal{F}_t$ gives the same as conditioning on the current state $S(t)$. Furthermore, $S(t)$ is a time-homogeneous Markov process since the distribution of the increment of Brownian motion only depends on $s - t$, and not on s and t separately. The reader should note that by substituting α with r, the risk-free return from the bond, geometric Brownian motion remains a Markov process when we change the probability to $\mathcal{Q}$.

Let $X = f(S(T))$ be the payoff from a contingent claim. From the pricing theory of general claims we know that

$$P(t) = e^{-r(T-t)} \mathbb{E}_\mathcal{Q}\left[f(S(T)) \mid \mathcal{F}_t\right],$$

and using the Markov property of $S(t)$ with respect to $\mathcal{Q}$ gives

$$P(t) = e^{-r(T-t)} g(t, T, S^{0,S(0)}(t)) = C(t, S(t)),$$

with $g(t, T, y) = \mathbb{E}_\mathcal{Q}\left[f(S^{t,y}(T))\right]$. In the last equality we used Thm. 4.11. Hence, we showed the link between the general pricing formula for contingent claims in Thm. 4.13 and the formula in Thm. 4.11 for special claims $X = f(S(T))$.

4.7 Contingent Claims on Many Underlying Stocks

Many options are based on more than one underlying stock. Examples are spread options and options on portofolios (called basket options). A typical spread option has payoff function $X = \max(S_1(T) - S_2(T), 0)$, where S_1 and S_2 are two stocks. A portfolio can be insured by buying a put option on the portfolio, which then has the payoff $X = \max(K - \sum_{i=1}^n a_i S_i(T), 0)$, where a_i is the number of shares held in stock i. A more exotic example of a multi-dimensional option can be a call option on the maximum of two stocks, having a payoff $X = \max(\max(S_1(T), S_2(T)) - K, 0)$. Our theory for pricing and hedging of contingent claims can easily be extended to options on many underlying stocks, which we now will show.

Assume our market consists of n stocks with price dynamics modelled by a multi-dimensional geometric Brownian motion, that is, for $i = 1, \ldots, n$, the price of stock i, S_i has dynamics

$$dS_i(t) = \alpha_i S_i(t) \, dt + S_i(t) \sum_{j=1}^n \sigma_{ij} \, dB_j(t), \tag{4.28}$$

where $B_1(t), \ldots, B_n(t)$ are n independent Brownian motions and α_i, σ_{ij} are constants. From (3.24) we know the explicit expression of S_i as

84   4 Pricing and Hedging of Contingent Claims

$$S_i(t) = S_i(0) \exp\left(\left(\alpha_i - \frac{1}{2}\sum_{j=1}^{n}\sigma_{ij}^2\right)t + \sum_{j=1}^{n}\sigma_{ij}B_j(t)\right).$$

Note that we have the same number of independent Brownian motions as different stocks. The market is assumed to have a risk-free investment opportunity modelled as before by a bond yielding a continuously compounding rate of return r.

The first we ask for is the existence of an equivalent martingale measure $\mathcal{Q}$ in this market. When we have many risky investment opportunities, $\mathcal{Q}$ is a martingale probability whenever the discounted price process of *each* stock is a martingale with respect to $\mathcal{Q}$. We construct $\mathcal{Q}$.

Let Σ denote the volatility matrix

$$\Sigma = \begin{bmatrix} \sigma_{11}, & \ldots, & \sigma_{1n} \\ \cdot & \cdot \cdot & \cdot \\ \cdot & \cdot \cdot & \cdot \\ \sigma_{n1}, & \ldots, & \sigma_{nn} \end{bmatrix},$$

and assume that Σ is non-singular so that its inverse Σ^{-1} exists. Introduce the column vector $\lambda := \Sigma^{-1}(\alpha - r\mathbf{1}) \in \mathbb{R}^n$ with $\alpha = (\alpha_1, \ldots, \alpha_n)'$ and $\mathbf{1}$ is the n-dimensional column vector of ones. Define the stochastic process $M(t)$ as

$$M(t) = \exp\left(-\lambda'\mathbf{B}(t) - \frac{1}{2}\lambda'\lambda t\right), \tag{4.29}$$

where $\mathbf{B}(t) = (B_1(t), \ldots, B_n(t))'$. For every $A \subset \Omega$, define the probability measure $\mathcal{Q}$ by

$$\mathcal{Q}(A) = \mathbb{E}\left[1_A M(T)\right]. \tag{4.30}$$

We have that $\mathcal{Q}$ is equivalent with $\mathcal{P}$ by construction. But is every discounted price dynamics a martingale with respect to $\mathcal{Q}$?

From the multi-dimensional version of Girsanov's theorem (see, e.g. [33, 41]), the stochastic process $\mathbf{W}(t)$

$$\mathbf{W}(t) := \mathbf{B}(t) + \lambda t, \tag{4.31}$$

defines an n-dimensional Brownian motion with respect to $\mathcal{Q}$. In Exercise 4.20 we prove that the stock price dynamics with respect to $\mathcal{Q}$ becomes

$$dS_i(t) = rS_i(t)\,dt + S_i(t)\sum_{j=1}^{n}\sigma_{ij}\,dW_j(t), \tag{4.32}$$

for $i = 1, \ldots, n$, and hence from the Itô formula we see that

$$d(e^{-rt}S_i(t)) = e^{-rt}S_i(t)\sum_{j=1}^{n}\sigma_{ij}\,dW_j(t). \tag{4.33}$$

4.7 Contingent Claims on Many Underlying Stocks

Thus, the discounted stock prices are sums of Itô integrals which should imply the martingale property. However, we must be a bit careful here, since we have not precisely stated what one means by a conditional expectation when more than one Brownian motion generates the information. The extension of $\mathcal{F}_s$ to multi-dimensions is straightforward. We let $\mathcal{F}_s$ be the collection of all sets $A \subset \Omega$ generated by *all* the Brownian motions $B_1(u), \ldots, B_n(u)$ for $u \leq s$. Since each stock price is made of all the n Brownian motions, $\mathcal{F}_s$ contains all the information generated by the stock prices up to time s. It is not a surprise that the Martingale representation theorem, Thm. 3.12, also holds in this case, and we find that $S_i^{\mathrm{d}}(t) := \mathrm{e}^{-rt} S_i(t)$ is a martingale, i.e.

$$\mathbb{E}_\mathcal{Q}\left[S_i^{\mathrm{d}}(t) \mid \mathcal{F}_s\right] = S_i^{\mathrm{d}}(s),$$

for every $i = 1, \ldots, n$ and $0 \leq s \leq t$. In conclusion, we have shown that $\mathcal{Q}$ is an equivalent martingale measure. In fact, it is the only such probability which is equivalent to $\mathcal{P}$ and turns all the discounted stock prices into martingales. A contingent T-claim is a contract where the owner receives the payoff X at time T, where X is a random variable which is $\mathcal{F}_T$-adapted.

We now have all the necessary ingredients to start the derivation of the price of contingent claims X in the multi-dimensional setting. As in the situation with only one underlying stock, we assume that the market is complete and does not allow for any arbitrage opportunities. The completeness provides us with the existence of a replicating portfolio consisting of all the n risky stocks and the bond. The value of the hedging portfolio at time $t \leq T$, is

$$H(t) = \sum_{i=1}^{n} a_i^H(t) S_i(t) + b^H(t) R(t),$$

where $a_i^H(t)$ is the number of shares in stock i at time t, while $b^H(t)$ is the number of bonds. Self-financing means

$$\mathrm{d}H(t) = \sum_{i=1}^{n} a_i^H(t) \, \mathrm{d}S_i(t) + b^H(t) \, \mathrm{d}R(t).$$

Now, since S_i^{d} is a martingale with respect to $\mathcal{Q}$, we have as before that also $\mathrm{e}^{-rt} H(t)$ is a martingale with respect to $\mathcal{Q}$. From the absence of arbitrage assumption it follows that the price $P(t)$ at time t of the claim X must be equal to the value of the hedging portfolio. Hence, $\mathrm{e}^{-rt} P(t) = \mathrm{e}^{-rt} H(t)$, and appealing to the martingale property and the fact that $H(T) = X$, we find

$$P(t) = \mathrm{e}^{-r(T-t)} \mathbb{E}_\mathcal{Q}\left[X \mid \mathcal{F}_t\right]. \tag{4.34}$$

We recognize this formula from the one-dimensional case; however, the definitions of $\mathcal{Q}$ and $\mathcal{F}_t$ are slightly different.

Considering claims of the form $X = f(S_1(T), \ldots, S_n(T))$, an analogous derivation to the one in Sect. 4.3 gives

$$P(t) = C(t, S_1(t), \ldots, S_n(t)), \tag{4.35}$$

where $C(t, x_1, \ldots, x_n)$ solves the multi-dimensional Black & Scholes partial differential equation

$$\frac{\partial C}{\partial t} + r \sum_{i=1}^{n} x_i \frac{\partial C}{\partial x_i} + \frac{1}{2} \sum_{i,j=1}^{n} \left(\Sigma \Sigma' \right)_{ij} \frac{\partial^2 C}{\partial x_i \partial x_j} C - rC = 0, \tag{4.36}$$

with terminal condition $C(T, x_1, \ldots, x_n) = f(x_1, \ldots, x_n)$. By $(\Sigma \Sigma')_{ij}$ we mean the element in row i and column j of the matrix obtained after multiplying Σ by its transpose. The solution can be represented as

$$C(t, x_1, \ldots, x_n) = e^{-r(T-t)} \mathbb{E}_Q \left[f(S_1^{t,x_1}(T), \ldots, S_n^{t,x_n}(T)) \right], \tag{4.37}$$

where $S_i^{t,x_i}(T)$ is given as

$$S_i^{t,x_i}(T) = x_i \exp \left(\left(r - \frac{1}{2} \sum_{j=1}^{n} \sigma_{ij}^2 \right) (T-t) + \sum_{j=1}^{n} \sigma_{ij} \left(W_j(T) - W_j(t) \right) \right).$$

The replicating portfolio is given by $a_i^H(t) = \partial C(t, S_1(t), \ldots, S_n(t))/\partial x_i$. This ends what we want to say about claims written on many underlying stocks.

4.8 Completeness, Arbitrage and Equivalent Martingale Measures

Throughout this chapter we have assumed that our markets are arbitrage-free and complete. It turns out, however, that both the market with one stock and the one with many stocks defined above *are* complete and arbitrage-free and the assumption is superfluous. The reason for this is a very deep connection between arbitrage, completeness and the existence of equivalent martingale measures, which is known as the *fundamental theorem of asset pricing*. Next to the Black & Scholes formula, this result constitutes one of the major achievements of mathematical finance.

Theorem 4.15. *The market does not allow for any arbitrage opportunities if and only if there exists at least one equivalent martingale measure Q. If there exists only one equivalent martingale measure, the market is in addition complete.*

For a proof of a very general version of the fundamental asset pricing theorem and references to the history of this result, we refer to the paper by Delbaen and Schachermayer [18].

Having an equivalent martingale measure we can price a claim using the conditional expectation with respect to this probability. We know that the

4.8 Completeness, Arbitrage and Equivalent Martingale Measures

price dynamics of the claim then becomes a martingale after discounting. Moreover, any self-financing portfolio also is a martingale after discounting, so the existence of an equivalent martingale measure prohibits arbitrage possibilities. One can obtain a pretty good insight behind the strong connection of arbitrage, completeness and equivalent martingale measures by going to discrete-time markets (see, e.g. [44]). However, the proof of Thm. 4.15 is very difficult and not suitable for the level of this book. We emphazise that Thm. 4.15 only holds when we assume no frictions in the market like transaction costs or portfolio constraints.

We can relate the existence and uniqueness of an equivalent martingale measure to the number of independent Brownian motions and different stocks for the multi-dimensional geometric Brownian motion (3.23).

Theorem 4.16. *Assume that the stocks are modelled by the dynamics in (3.23) (or, equivalently, (3.24)). Then,*

1. *The market is arbitrage-free if and only if $n \leq m$,*
2. *The market is complete if and only if $n \geq m$.*

In Sects. 4.3 and 4.5, we modelled the price dynamics of the underlying stock with a geometric Brownian motion where $n = m = 1$. Hence, this market is both complete and arbitrage-free. In the multi-dimensional case considered in Sect. 4.7, we assumed $n = m$ different stocks and Brownian motions, which again yields a complete and arbitrage-free market.

The completeness comes from the fact that we need to be able to trade in every source of noise, here represented by different Brownian motions. Hence, we must at least have the same number of linearly independent stocks as we have Brownian motions. The existence of equivalent martingale measures is linked to the well-definedness of $M(T)$, which depends on the possibility to invert Σ. The matrix Σ is non-singular exactly when $n = m$ (being quadratic) and the stocks are linearly independent (giving linearly independent rows of Σ).

Theorem 4.15 holds in markets where the stocks have much more general dynamics than (multi-dimensional) geometric Brownian motion. By a slightly more restrictive interpretation of arbitrage,[8] it covers markets where the stocks are modelled by Lévy processes like in Sect. 2.6. As a matter of fact, even Thm. 4.16 can be used for such markets, giving a rule of thumb for when to expect that the market under consideration is complete and free of arbitrage. For instance, if we assume we have one stock with the exponential NIG Lévy process as pricing dynamics, we apparently have only *one* source of noise. We could believe that this defines a complete and arbitrage-free market. However, the market will *not* be complete since the Lévy process introduces noise that cannot be traded. As a matter of fact, the NIG Lévy

---

[8] The stronger notion of arbitrage used is called "No free lunch with vanishing risk".

88   4 Pricing and Hedging of Contingent Claims

process introduces an *infinite* amount of noise sources. The process jumps at random times. When it jumps, it jumps at a random size. For each jumpsize, we have one source of noise, and since the NIG Lévy process can jump at all sizes, we have uncountably many sources of noise. We therefore in principle need a continuum of stocks to trade in to have a complete market. Equivalent martingale measures can, however, be constructed, so the market is at least arbitrage-free.

We also make some remarks concerning stochastic volatility models. It is quite popular to let the stock price be described by a geometric Brownian dynamics where σ is a stochastic process being a semimartingale:

$$dS(t) = \alpha S(t)\,dt + \sigma(t) S(t)\,dB(t),$$
$$d\sigma(t) = u(t,\sigma(t))\,dt + v(t,\sigma(t))\,d\widetilde{B}(t).$$

The stochastic volatility process $\sigma(t)$ is driven by a Brownian motion $\widetilde{B}(t)$ correlated with $B(t)$. We end up with a market where there are two Brownian motions (although correlated, they are far from being identical), and only one stock, and hence we have an incomplete market model. In this case we cannot expect to be able to replicate even plain vanilla options like calls and puts.

When the market is incomplete, we have many equivalent martingale measures, which raises the question of what should be the price of a claim. If we cannot hedge it, what is then the arbitrage-free price? Indeed, if the market is arbitrage-free, we have many equivalent martingale measures, which in turn leads to many suggested prices via the conditional expectation of the claim payoff with respect to each of these probabilites. Thus, we have many arbitrage-free prices to choose among. How to deal with this problem is the topic of the next section.

4.9 Extensions to Incomplete Markets

In the Black & Scholes model there are a lot of implicit assumptions about the market mechanism. For instance, it is supposed that an investor can trade without paying any transaction costs, which is a highly unrealistic assumption. Indeed, this is critical in the derivation of the arbitrage-free price of a claim, since we must incessantly rebalance our hedging portfolio. Recalling that the number of shares of the underlying stock is $a^H(t) = \partial C(t, S(t))/\partial x$ for a claim $X = f(S(T))$, we see that the hedge is continuously updated. If the market charges transaction costs, such a strategy would become infinitely costly and therefore not reasonable. On the other hand, this is the only way to replicate the claim, and therefore the market will become incomplete in practice.

There exist other market frictions as well. For instance, regulators are usually enforcing constraints on how short investors can go in stocks and bonds. Furthermore, in many situations the market can be very illiquid in the

4.9 Extensions to Incomplete Markets

sense that you do not find a buyer for your stock when you want to sell, and vice versa. The most extreme is the market for options on weather, where you cannot trade at all in the underlying "stock" (you cannot buy temperature, for instance). In the Black & Scholes setup these market frictions are assumed not to exist. Imposing one or more of these frictions will inevitably lead to an incomplete market.

Another aspect that we have already touched on earlier is incompleteness due to stock price dynamics which deviates from geometric Brownian motion. For example, if the stock price is modelled by a geometric NIG Lévy process or by some stochastic volatility model, the market becomes incomplete. The degree of incompleteness increases even more if we add transaction costs and constraints on the hedging positions.

Incompleteness is a part of financial markets, and it is impossible to avoid in practice (although one can have approximate completeness in some circumstances). If we have had completeness in the markets for derivatives, options and claims would not exist simply because they would be redundant. We could achieve exactly the same by entering into the claim's replicating portfolio. Today's derivatives markets are huge, and the reason is obviously that options introduce genuinely new investment opportunities. Market frictions and deviations from normality in stock returns force us to take into account the non–hedgeability of claims. The natural question is therefore: what is the price of a claim when the market is incomplete? In the mathematical finance literature there exist many techniques in order to resolve the pricing problem, and we will in the rest of this section briefly mention a few of them.

In general it is impossible to find a replicating portfolio for a claim when the underlying stock is modelled as a geometric Lévy process. A contingent claim therefore introduces a new investment opportunity, and potentially also an arbitrage opportunity. If we do not allow for arbitrage in the extended market consisting of the bond, underlying stock and the claim, we have a starting point for deriving the price of the claim. As we have seen in Sect. 4.8, the absence of arbitrage is equivalent to existence of a risk-neutral probability measure $\mathcal{Q}$, and the price of the claim is the discounted expected payoff conditioned on the information up to time t. Here the conditional expectation is taken with respect to the risk-neutral probability. With the underlying stock modelled by a geometric Lévy process, we are in the awkward situation of having an abundance of risk-neutral measures $\mathcal{Q}$, each giving an arbitrage-free pricing dynamics for the claim. If we look at the prices at time 0,

$$P(0) = e^{-rT}\mathbb{E}_{\mathcal{Q}}[X],$$

we obtain an interval of prices when calculating the expectation for all different risk-neutral measures $\mathcal{Q}$. Our problem is to select one of the prices $P(0) \in [m, M]$, where

$$m = \inf\left\{e^{-rT}\mathbb{E}_{\mathcal{Q}}[X] \mid \mathcal{Q} \text{ equivalent martingale measure}\right\},$$
$$M = \sup\left\{e^{-rT}\mathbb{E}_{\mathcal{Q}}[X] \mid \mathcal{Q} \text{ equivalent martingale measure}\right\}.$$

Usually this interval is large. For example, Eberlein and Jacod [21] demonstrate that for claims $X = f(S(T))$ on a stock with price dynamics modelled as a geometric Lévy process the interval will have limits $m = e^{-rT} f(e^{rT} S(0))$ and $M = S(0)$. For a call option this means $m = \max(S(0) - e^{-rT}K, 0)$. It is possible to show that if the price is outside of this interval, one can obtain arbitrage. Hence, a priori we know that the arbitrage-free prices must be between m and M. The argument is independent of the model for the underlying stock, and the reader is encouraged to work this out in Exercise 4.4. In [21] is shown that *all* prices between m and M are abitrage-free. Moreover, there is a nice interpretation of m and M in terms of sub- and superhedging strategies, which we now discuss in a slightly different context.

In markets where there are costs connected with transactions and other frictions like short-selling constraints, there will also exist an interval of arbitrage-free prices $P_0 \in [m, M]$. The lower and upper limits of the interval are related to sub- and superhedging strategies, resp. In fact, when we impose market frictions into our model we do not talk any longer about equivalent martingale measures, but instead derive the pricing interval as the cost to subreplicate or superreplicate the claim. A superhedging strategy is a self-financing portfolio consisting of a position in the bond and the underlying stock which at the time of exercise pays *at least* the value of the claim. A subhedging strategy pays, on the other hand, *at most* the payoff from the claim. The value of m is the price of the most expensive subhedging strategy, while M is the price of the cheapest superhedging strategy. To state this in a mathematical language, let $H_{\text{sub}}(t)$ be the value of a subhedging portfolio of a claim X at time t. Then, $H_{\text{sub}}(T) \leq X$. We define

$$m = \sup \{H_{\text{sub}}(0) \mid H_{\text{sub}}(t) \text{ subhedging portfolio}\}.$$

A superhedging portfolio $H_{\text{super}}(t)$ has the property $H_{\text{super}}(T) \geq X$, and M is defined analogously to m, i.e.,

$$M = \inf \{H_{\text{super}}(0) \mid H_{\text{super}}(t) \text{ superhedging portfolio}\}.$$

The theory of sub- and superhedging strategies in markets with constraints is presented in detail by Karatzas and Shreve [34].

The literature suggests many ways to pick the "right" price $P(0)$ from the arbitrage-free pricing interval. The approaches differ with the type of incompleteness we have. We are now going to sketch some of the ideas to resolve the pricing problem in an incomplete market, and give the reference to where the reader can go into more details.

If the market we consider incurs proportional transaction costs, Hodges and Neuberger [29] suggest to use ideas from portfolio optimization theory to derive a fair price on call options. In the setup of Hodges and Neuberger, investors pay a proportion of the total trade in transaction cost every time they decide to rebalance their portfolio. The authors consider the following problem for *issuers* of a call option: they can decide *not* to issue the option, but instead optimize the utility of their wealth w by investing in the underlying stock and bond. Otherwise, they can issue the option and receive a premium $P(0)$ which they can add to their wealth and optimize the expected utility of their investments. At the time of exercise, they must settle the claim from the buyer of the option. The fair premium for the option is now the price $P(0)$ that make the issuers *indifferent* between the two investment alternatives, that is, the premium that gives the same expected utility. Interestingly, when transaction costs go to zero, the fair premium $P(0)$ becomes simply the Black & Scholes price. This way of pricing options has been generalized by Davis, Panas and Zariphopoulou [17], among others.

If the underlying stock of the claim follows a geometric Lévy process, Gerber and Shiu [28] suggest to use the *Esscher transform* to derive explicitly one risk-neutral probability Q. The Esscher transform is widely used in insurance mathematics, and works well for many Lévy processes. It can be connected with the risk aversion of an investor, thereby linking pricing of claims in incomplete markets to the risk preferences of the investor (see discussion in [28]). Chan [12] studies the pricing for such stock price models from a more theoretical perspective. He considers equivalent martingale measures Q which in some sense are closest to the unknown measure chosen by the market. Among other things, [12] proves that the Esscher transform defines the so-called entropy probability. Föllmer and Schweizer [26] construct the entropy probability from a minimization criterion related to the claim X. Another, more statistical approach to fixing the equivalent martingale measure Q is to look at historical option prices. Eberlein, Keller and Prause [23] parametrize the class of equivalent martingale measures Q and fit the theoretically derived prices to the observed historical ones, and in this way estimate the equivalent martingale measure chosen by the market. Their approach is motivated by stochastic interest rate theory where one uses a similar technique to fit the observed yield curve.

Even though the pricing problem in incomplete markets can be solved using highly sophisticated mathematical techniques, the Black & Scholes methodology provides in many circumstances a very good approximation to the "right" price. The explicit formula for options together with easily accessible theory makes it attractive for practitioners despite its many shortcomings.

4.9.1 Energy Markets and Incompleteness

In this subsection we shall investigate in more detail the consequences of incompleteness on arbitrage-free prices. Energy markets are very relevant in this respect. Consider for instance the spot market for electricity[9] where you

---

[9] There are many of them in the US and Europe. The first liberalized market for electricity was the NordPool electricity exchange in Scandinavia. It started in 1993 in Oslo, Norway.

can reade not only in electricity produced in the Scandinavian countries, but also in derivatives like call and put options with the spot as the underlying "stock". If we bluntly assume that spot electricity follows a geometric Brownian motion[10]

$$dS(t) = \alpha S(t)\,dt + \sigma S(t)\,dB(t),$$

and the risk-free investment alternative is a bond with constant rate of return r,

$$dR(t) = rR(t)\,dt,$$

we are back to the Black & Scholes market. Apparently, we have a complete market where we can price options and other derivatives using the arbitrage-free pricing technology which we have discussed in the previous sections. Unfortunately, this is not possible. If you buy electricity on the spot market, it is far from being the same as buying shares in some company quoted on a stock exchange. Electricity is a physical product, unlike a stock which in most exchanges is simply virtual paper existing in some computer; it must be used once you have bought it. You cannot, of course, store electricity and therefore it is not suitable for hedging of an option.[11]

The electricity market is incomplete, because we cannot use the spot for hedging. We are left with the bond as the only financial instrument to hedge a claim, which, of course, is not possible. Ordinary call and put options are therefore not hedgeable. We can see this from the perspective of equivalent martingale measures. When we define an equivalent martingale measure in Def. 4.9, we effectively say that *all* tradeable assets should be martingales with respect to the probability $\mathcal{Q}$ after discounting in order for this to be a martingale probability. We interpret as *tradeable* those assets which can be used in hedging. The spot electricity is not among them. Discounting the bond leaves us with simply a constant 1, which is trivially a martingale with respect to all probabilites $\mathcal{Q}$ equivalent to $\mathcal{P}$. Hence, all probabilities equivalent to $\mathcal{P}$ will be equivalent martingale measures.

Girsanov's theorem (Thm. 4.10) produces probabilities $\mathcal{Q}$ which are equivalent to $\mathcal{P}$ *and* for which a certain process becomes a Brownian motion. Let $\lambda \in \mathbb{R}$ be any constant; we see from Thm. 4.10 that $\mathcal{Q}$ defined by $Q(A) = \mathbb{E}\left[1_A M(T)\right]$ for $A \subset \Omega$ and

$$M(T) = \exp\left(-\lambda B(T) - \frac{1}{2}\lambda^2 T\right),$$

becomes an equivalent probability of $\mathcal{P}$. Furthermore, the process

---

[10] This is not a realistic model for the electricity dynamics. See, for example [24] and [6] for discussion on different types of stochastic processes modelling energy dynamics.

[11] If electricity is produced by hydro power, you can store it indirectly in a reservoir. However, this is a very expensive hedging approach.

4.9 Extensions to Incomplete Markets

$$W(t) = B(t) + \lambda t,$$

is a Brownian motion with respect to this probabilty. Hence, we have a collection of *equivalent martingale measures* (or risk-neural measures), one for each choice of λ.

Let us see how we can find arbitrage-free prices for a call option on the spot electricity with strike K at exercise time T. From the theory an arbitrage-free price of the call option is given by

$$P(0) = e^{-rT} \mathbb{E}_{\mathcal{Q}}\left[\max(S(T) - K, 0)\right],$$

where $\mathcal{Q}$ is an equivalent martingale measure. Specify a λ, and we have fixed a $\mathcal{Q}$ which we now denote by $\mathcal{Q}(\lambda)$. We find the dynamics of $S(t)$ with respect to $\mathcal{Q}(\lambda)$,

$$\begin{aligned}
dS(t) &= \alpha S(t)\,dt + \sigma S(t)\,dB(t) \\
&= \alpha S(t)\,dt + \sigma S(t)\,(dW(t) - \lambda dt) \\
&= (\alpha - \sigma\lambda)S(t)\,dt + \sigma S(t)\,dW(t),
\end{aligned}$$

where $W(t)$ is a Brownian motion with respect to $\mathcal{Q}(\lambda)$. The reader is now asked in Exercise 4.22 to derive the pricing formula

$$P(0) = e^{-(r-\alpha+\sigma\lambda)T} S(0)\Phi(d_1) - K e^{-rT}\Phi(d_2), \tag{4.38}$$

with $d_1 = d_2 + \sigma\sqrt{T}$ and

$$d_2 = \frac{\ln(S(0)/K) + (\alpha - \sigma\lambda - \sigma^2/2)T}{\sigma\sqrt{T}}.$$

Here the price $P(0)$ will depend on the choice of λ, a parameter which is usually referred to as *the market price of risk*. It measures the additional premium charged by the issuer of the option for taking the unhedgeable risk of entering a call option on electricity. This parameter can be fixed, for instance, by looking at historical price quotes of options.

But what can we say about the arbitrage-free pricing interval? If we let $\lambda \to +\infty$, d_1 and d_2 will both tend to $-\infty$. Hence, $\Phi(d_1)$ and $\Phi(d_2)$ tend to zero, and since $\exp(-\sigma\lambda T)$ tends to zero as well, the price $P(0)$ will converge to zero. The lower arbitrage-free price is thus zero. But can we find any upper bound? Letting $\lambda \to -\infty$, it is easy to show that $d_1, d_2 \to \infty$ and $\Phi(d_1), \Phi(d_2)$ tend to 1. But $\exp(-\sigma\lambda T) \to \infty$, and hence $P(0) \to \infty$. Thus, we conclude that the arbitrage-free pricing interval for electricity calls is $(0, \infty)$.

We end with a few comments on the market price of risk λ. It is a common approach in electricity markets (and more generally in energy markets) to find λ by looking at historical forward price curves. One can derive theoretical forward prices on electricity, and then try to find the λ that best fits theoretical forward prices to the observed ones. The fitted λ can now be used for pricing of options and other derivatives on electricity (see, for instance, [6] for more on this approach). The market price of risk reflects the trader's view on exposing themselves to risk. Thus, it is initmately connected with the risk preferences of the market participants, thereby motivating a utility maximization approach for option pricing. Davis [16] used this method in another highly incomplete market with the same hedging problem as for spot electricity, namely the market for temperature derivatives.

Exercises

4.1 In the one-period market defined in Sect. 4.1, show how one can construct arbitrage when the price of a claim X is $\tilde{P} > P(0)$ in the market. We denote by $P(0)$ the arbitrage-free price.

4.2 Consider a one-period market with stock and bond. We suppose the stock has *three* different possible outcomes at time T, that is, with $\Omega = \{\omega_1, \omega_2, \omega_3\}$ we assume $S(T, \omega_i) = s_i$, for $i = 1, 2, 3$, and $s_1 > s_2 > s_3$. Show that it is *not* possible in general to hedge a claim X in this market.

4.3 Assume we are in a one-period market where the stock initially is s and has three outcomes at time T given by $S(T, \omega_1) = sd$, $S(T, \omega_2) = s$ and $S(T, \omega_3) = su$ where $0 < d < 1 < u$. We assume that the bond pays zero interest, $r = 0$. Consider a contingent claim paying x if the stock goes up, and y if it is unchanged, while it pays zero when the stock goes down.

  a) Characterize all portfolios which have *at least* the value of the payoff from the claim at time T. What is the price of the cheapest such portfolio? We call such portfolios *superreplicating* portfolios.

  b) Do the same analysis for so-called *subreplicating* portfolios, that is, those portfolios which have *at most* the value of the payoff from the claim at time T.

  c) Show that all prices of the claim being greater than most expensive subhedge, and smaller than the cheapest superhedge, will not allow for any arbitrage opportunities.

4.4 Let $P(t)$ be the price of a call option at time t, where K is the strike price and $T > t$ the time of exercise. Show that

$$P(t) \geq S(t) - Ke^{-rt},$$

otherwise there exists an arbitrage opportunity. Argue by arbitrage that $P(t) \leq S(t)$.

4.5 Let $H(t)$ be the value of the replicating portfolio of a contingent claim X in a Black & Scholes market. Show that if the claim is traded for a price $\tilde{P}(t) > H(t)$, then there exist arbitrage opportunities.

4.6 Consider the function $p(t, z) = (1/\sqrt{2\pi t}) \exp(-z^2/2t)$. Show that

$$\frac{\partial p(t, z)}{\partial t} = \frac{1}{2} \frac{\partial^2 p(t, z)}{\partial z^2}.$$

4.9 Extensions to Incomplete Markets

4.7 Consider the process $Z^{t,x}(s)$ defined in Thm. 4.4. Find the density function $p(t,z)$ for $\ln Z^{t,x}(T)$. Show that

$$C(t,x) = e^{-r(T-t)} \int_{-\infty}^{\infty} f(e^z) p(t,z) \, dz,$$

and use this to prove Thm. 4.4.

4.8 From the Black & Scholes formula for a call option with strike K and exercise time T we know that arbitrage-free price is a function of T, K, risk-free interest rate r, initial stock price $S(t)$ and volatility σ. Denote the price at time t as $P(t;T,K,r,S(t),\sigma)$, and find

$$\lim_{\sigma \downarrow 0} P(t;T,K,r,S(t),\sigma), \quad \lim_{\sigma \uparrow \infty} P(t;T,K,r,S(t),\sigma).$$

Show that $P(t;T,K,r,S(t),\sigma)$ is an increasing function in T and σ, and decreasing in K.

4.9 Find the arbitrage-free price of a *digital option* with strike $K = S(0)$ (today's stock price) and exercise time T. The digital option pays €1 if $S(T) > S(0)$ at the time of exercise, and zero otherwise.

4.10 Derive the put–call parity,

$$P_t^c - P_t^p = S_t - Ke^{-r(T-t)},$$

where P_t^c is the price of a call option and P_t^p is the price of a put option. Both have exercise price K at maturity T. Hint: use $\max(x-K,0) = (x-K) + \max(K-x,0)$.

4.11 Find the hedging strategy (or the delta) for a put option.

4.12 Use the density method to calculate the hedge of a digital option which pays €1 when the price of the underlying stock is greater than K at time of exercise T, and zero otherwise.

4.13 Show that $\mathcal{Q}$ defined in (4.18) is a probability.

4.14 Show that $S_t^d = e^{-rt} S_t$ is a martingale with respect to the probability $\mathcal{Q}$ defined in (4.18).

4.15 Let $M(t) = \exp\left(-\lambda B(t) - \frac{1}{2}\lambda^2 t\right)$ for a constant $\lambda \in \mathbb{R}$. Show that $M(t)$ is a martingale.

4.16 Let $\mathcal{Q}$ be defined as in (4.18), and consider the stochastic process $W(t) = B(t) + \lambda t$. Calculate the moment generating function of $W(t)$ with respect to the probability $\mathcal{Q}$. In the exercise, you are not supposed to appeal to Girsanov's theorem, which will immediately give you the moment generating function (Exercise 3.10).

4.17 Let (a,b) be a self-financing trading strategy. If $H(t)$ is the value at time t of this portfolio, show that the discounted portfolio value is a martingale with respect to the equivalent martingale measure $\mathcal{Q}$ of the Black & Scholes market.

4.18
   a) From (4.25), derive a Black & Scholes-type formula for the chooser option.
   b) Prove that the price of the chooser option also can be written as
   $$P(0) = P^{\mathrm{p}}(0) + e^{-rt}\mathbb{E}_\mathcal{Q}\left[\max\left(0, S(t) - Ke^{-r(T-t)}\right)\right],$$
   where $P^{\mathrm{p}}(0)$ is the price of a put option with strike K at exercise time T.

4.19 A *forward contract* on a stock is a financial agreement that guarantees the owner delivery of one share at an agreed time T. The owner pays a contracted price for the share. If the contract is entered into at time $t < T$, the so-called *forward price*, denoted by $F(t,T)$, is defined to be the price the owner pays to the counterpart for the share at time of delivery. Note that this contract looks very similar to a call option, however, with the difference that the owner is *obliged* to buy the share for the agreed price $F(t,T)$. We assume the Black & Scholes market.
   a) Show that the payoff function from the forward contract is $X = S(T) - F(t,T)$, and the arbitrage-free price is
   $$P(t) = e^{-r(T-t)}\mathbb{E}\left[S(T) - F(t,T)\,|\,\mathcal{F}_t\right].$$
   b) A distinctive feature with the forward contract is that $F(t,T)$ is chosen such that the arbitrage-free price is 0 at time of entry. Based on this one can derive the forward price explicitly: when entering a forward contract at time t, one only has information about the stock price up to time t, and it is therefore natural to assume that $F(t,T)$ is $\mathcal{F}_t$-adapted since it can only be settled purely based on the available information. Find $F(t,T)$ such that $P(t) = 0$.
   c) Derive the dynamics of $F(t,T)$ (as a stochastic process of t) with respect to the risk-neutral probability $\mathcal{Q}$.
   d) Derive a Black & Scholes formula for the arbitrage-free price of a call option with strike K and exercise τ written on a forward contract with delivery at time $T > \tau$.

4.20 Show that the multi-dimensional stock price defined in (4.28) has dynamics (4.32) with respect to the equivalent martingale measure $\mathcal{Q}$ for the choice $\lambda = \Sigma^{-1}(\alpha - r\mathbf{1})$.

4.21 Consider the two-dimensional geometric Brownian motion
$$dS_1(t) = \alpha_1 S_1(t)\,dt + \sigma_1 S_1(t)\left\{dB_1(t) + \rho dB_2(t)\right\},$$
$$dS_2(t) = \alpha_2 S_2(t)\,dt + \sigma_2 S_2(t)dB_2(t),$$

where $\alpha_i, \sigma_i > 0$ and ρ are constants, $i = 1, 2$, and B_1, B_2 are two independent Brownian motions (see Exercise 3.14 for a discussion of such a market model).

a) Find the equivalent martingale measure Q in the market consisting of two stocks with price dynamics described by $S_1(t)$ and $S_2(t)$, and a bond paying r in risk-free interest. Find $W_1(t)$ and $W_2(t)$, the two independent Brownian motions with respect to Q. What is the dynamics of the stocks with respect to this equivalent martingale measure?

b) A *spread option* is an option written on the difference of two stocks. Consider a call option on the difference of two stocks, with strike K and time of exercise T, which has payoff function

$$X = \max\left(S_1(T) - S_2(T) - K, 0\right).$$

In general[12] it is not possible to find an explicit pricing formula for this spread option. However, we can derive an approximation. Typically, the distribution of $S_1(T) - S_2(T)$ with respect to Q is close to normal. Use this information to find an approximate price $P(0)$ at time 0 for the spread option.

4.22 Derive the pricing formula (4.38) for call options on electricity.

---

[12] For $K = 0$, the price is given by the Margrabe formula [38].

5 Numerical Pricing and Hedging of Contingent Claims

This chapter introduces numerical methods for calculating prices and replication strategies of contingent claims. Only a few options can be priced and hedged explicitly, for instance, call and put options. For path-dependent derivatives like barrier and average (Asian) options there does not exist any explicit formula, and a numerical approach is necessary to evaluate them. We will introduce techniques based on Monte Carlo simulations of the risk-neutral expectation defining the derivatives price, and so-called finite-difference methods which numerically solve the Black & Scholes partial differential equation (4.12). Both approaches work for claims with payoff of the form $f(S(T))$, while the former is also applicable for more general claims.[1]

5.1 Pricing and Hedging with Monte Carlo Methods

The Monte Carlo approach to option pricing and hedging is a very useful and easily implemented method. It can be used for numerical evaluation of nearly all derivatives, but has the disadvantage of being computationally slow. In this section we will demonstrate the flexibility of the method by developing algorithms for the pricing of barrier and Asian options, along with more standard options having payoff of the form $f(S(T))$. In addition, we shall address the questions of hedging and numerical efficiency of the Monte Carlo technique.

The idea behind the Monte Carlo technique is rather simple: let us assume that we would like to calculate the expectation $\mathbb{E}[g(Y)]$, where g is some known function and Y is a random variable.[2] Imagine we have a recipe for simulating independent outcomes from Y on a computer. Then we can draw N independent outcomes $(y^1, \ldots, y^N)$ from Y and calculate the average $\frac{1}{N} \sum_{i=1}^{N} g(y^i)$. This number will be an approximation of the expectation $\mathbb{E}[g(Y)]$, i.e.,

---

[1] The reader interested in a broad treatment of numerical methods in financial problems is referred to Jäckel [32].

[2] We now talk generally about the Monte Carlo technique. In a few moments we will discuss how it works for options.

$$\mathbb{E}[g(Y)] \approx \frac{1}{N} \sum_{i=1}^{N} g(y^i).$$

To link the Monte Carlo method to pricing of options, recall that the fair value of a contingent claim with payoff $f(S(T))$ is given as $e^{-rT}\mathbb{E}_{\mathcal{Q}}[f(S(T))]$, where $S(T)$ is a lognormal random variable under the probability $\mathcal{Q}$. Since one can simulate outcomes from normal variables using a computer,[3] we can approximate this price using the Monte Carlo technique described above.

5.1.1 Pricing and Hedging of Contingent Claims with Payoff of the Form $f(S_T)$

From Sect. 4.3 we recall that a contingent claim with payoff $X = f(S(T))$ has the value $P(t) = C(t, S(t))$, at time t, where

$$C(t, x) = e^{-r(T-t)} \mathbb{E}_{\mathcal{Q}}\left[f(S^{t,x}(T))\right]. \tag{5.1}$$

With respect to $\mathcal{Q}$, the dynamics of $S^{t,x}(T)$ is a geometric Brownian motion with drift r and volatility σ, and initial state at time t given as $S^{t,x}(t) = x$. In other words,

$$S^{t,x}(T) = x \exp\left(\left(r - \frac{1}{2}\sigma^2\right)(T-t) + \sigma\left(W(T) - W(t)\right)\right).$$

Furthermore, the number of stocks in the replicating portfolio is $a^H(t) = \partial C(t, S(t))/\partial x$, while the number of bonds is given by $b^H(t) = (C(t, S(t)) - a(t))S(t))/R(t)$. Thus, we need to calculate the delta $\partial C(t, S(t))/\partial x$ in addition to the price $C(t, S(t))$ in order to characterize the hedging portfolio. Remark that $\partial C(t, x)/\partial x$ also measures the sensitivity of the option price with respect to the value of the underlying stock.

Inserting $S^{t,x}(T)$ into the expression for $C(t, x)$, we get

$$C(t, x) = e^{-r(T-t)} \tag{5.2}$$
$$\times \mathbb{E}_{\mathcal{Q}}\left[f\left(x \exp\left(\left(r - \frac{1}{2}\sigma^2\right)(T-t) + \sigma(W(T) - W(t))\right)\right)\right].$$

Recall that $W(t)$ is a Brownian motion with respect to the probability $\mathcal{Q}$. Hence, in (5.3) we find the expectation of a function of a normal random variable with mean 0 and variance $T - t$. To simplify the expression slightly, let Y be a *standard normal* random variable. Then $\sqrt{T-t}\,Y$ will become a normally distributed random variable with zero expectation and variance $T - t$, and thus being equal to $W(T) - W(t)$ in distribution. Given a routine for drawing independent normally distributed numbers, we can numerically approximate $C(t, x)$ for fixed t and x using the following algorithm:

---

[3] Standard routines for drawing independent and normally distributed random numbers exist in most mathematical and statistical software packages and Excel.

5.1 Pricing and Hedging with Monte Carlo Methods

Algorithm 1 *Monte Carlo algorithm for the option price.*

1. *Draw N independent outcomes from the random variable $Y \sim \mathcal{N}(0,1)$:*

$$(y^1, \ldots, y^N)$$

2. *Calculate*

$$e^{-r(T-t)} \frac{1}{N} \sum_{k=1}^{N} f\left(x \exp\left(\left(r - \frac{1}{2}\sigma^2\right)(T-t) + \sigma\sqrt{T-t}\, y^k\right)\right).$$

The resulting number from step 2 in the algorithm is an approximation of the expression for $C(t,x)$ in (5.3). How close this number is to the real $C(t,x)$ depends on the size of N. Later in this section we are going to discuss the accuracy of the Monte Carlo method as a function of N.

To implement this algorithm we need a technique for drawing independent outcomes from a normally distributed random variable. We will now describe how one can do this using Visual Basic, the macro programming language of Excel. In Visual Basic there exists a function for drawing independent outcomes from a *uniformly* distributed random variable U on $[0,1]$. In Exercise 5.1 the reader is asked to prove that $Y = \Phi^{-1}(U)$ is a standard normal random variable, where we recall that Φ is the cumulative standard normal distribution function

$$\Phi(x) = \frac{1}{\sqrt{2\pi}} \int_{-\infty}^{x} e^{-z^2/2}\, dz,$$

and Φ^{-1} is its inverse. Since Visual Basic already has implemented the inverse function of Φ, we have an easy recipe for generating outcomes from Y. The following two lines provide an outcome y from Y:

```
u=Rnd()
y=Application.WorksheetFunction.NormSInv(u)
```

The function Rnd() gives the outcome from U, while the function NormSInv(u) corresponds to $\Phi^{-1}(u)$ in Visual Basic. Using this little program repeatedly, we can find independent outcomes of Y. There exist other, more efficient methods for generating numbers from a standard normal distribution (see, for example [32, 46]). We remark that the computer is not able to generate truly uniformly distributed numbers. This is why one often refers to these numbers as *pseudo-random*.

Below we present the Visual Basic code to implement a function that calculates the price of an option using the Monte Carlo algorithm. The function is called OptionMC and uses as input the current stock price x, current time t, risk-free interest rate r, volatility σ, exercise time T and number of Monte Carlo replications N:

```
Function OptionMC(x, t, r, sigma, et, N)
    ' x=current stock price
    ' t=current time
    ' r=interest rate
    ' sigma=volatility
    ' et=time of exercise
    ' N=number of Monte Carlo replications

    p=0
    For k=1 To N
        u=Rnd()
        y=Application.WorksheetFunction.NormSInv(u)
        p=p+PayOff(x*Exp((r-0.5*sigma*sigma)*
            (et-t)+sigma*Sqr(et-t)*y))
    Next
    p = p/N
    OptionMC=Exp(-r*(et-t))*p
End Function
```

The code can be entered in Excel using the Visual Basic modules and then used in the spreadsheet like any other function. Note, however, that PayOff must be defined by the user as the payoff function of the option. We assume here that this function takes one input, namely the price of the underlying stock $S^{t,x}(T)$. However, in practice we can have more inputs, for example, the strike price if we price a call option. In that case it may be natural to have the strike price as additional input to the function OptionMC itself (indeed, it may be reasonable to change the name of the function also). We leave these details to the reader who wants to try out the methods on a computer.

The Monte Carlo algorithm for calculating the delta of the option is analogous to the pricing algorithm. Recalling the density approach in Subsect. 4.3.5, the delta $\partial C(t,x)/\partial x$ can be written as an expectation. We find a representation of $\partial C(t,x)/\partial x$ with respect to the equivalent martingale measure in Thm. 4.12 as

$$\frac{\partial C(t,x)}{\partial x} = e^{-r(T-t)} \mathbb{E}_Q \left[g(t,x,S^{t,x}(T)) \right],$$

with

$$g(t,x,s) = f(s) \frac{\ln s - \ln x - (r - \sigma^2/2)(T-t)}{x\sigma^2(T-t)}.$$

Observe that

5.1 Pricing and Hedging with Monte Carlo Methods

$$g(t,x,S^{t,x}(T)) = f(S^{t,x}(T))\frac{W(T)-W(t)}{x\sigma(T-t)}.$$

We can now use Algorithm 1 to simulate the delta by simply substituting the payoff f with the "modified payoff" g. The Visual Basic function `OptionMC` can be used with `PayOff` defined to be the function g instead. Thus, we can use a more sophisticated `OptionMC` as the core function which does all Monte Carlo evaluations, and by varying the contents of `PayOff` we can get the price or the delta as the output according to our interest.[4]

So-called *numerical differentiation* is an alternative approach for evaluation of delta. We can approximate $\partial C(t,x)/\partial x$ by finding the price of the claim for two different initial states of the underlying stock. The definition of the derivative of $C(t,x)$ with respect to x is

$$\frac{\partial C(t,x)}{\partial x} = \lim_{h \to 0} \frac{C(t,x+h) - C(t,x)}{h},$$

and we can therefore numerically approximate this with

$$\frac{\partial C(t,x)}{\partial x} \approx \frac{C(t,x+h) - C(t,x)}{h},$$

for a small h. The delta $\partial C(t,x)/\partial x$ can be simulated by first calculating $C(t,x+h)$ and then $C(t,x)$ using the Monte Carlo algorithm. After dividing the difference of these numerically calculated prices by h, we end up with an approximation of the delta.

The numerical differentiation approach apparently needs two simulations of the option price to find the delta. It therefore seems to be twice as costly to compute as the technique based on the density approach. The density approach requires implementation of one additional function, but once this is done, the efficiency in finding the delta is the same as for Monte Carlo simulating the price. However, we can speed up the numerical differentiation technique by using the same drawings of normally distributed random numbers. Instead of simulating a new sequence of y^k's for calculating $C(t,x+h)$, we can use the same as for the approximation of $C(t,x)$. This also significantly reduces the error. The trick is thus to create a new function in Visual Basic which simulates $C(t,x+h)$ and $C(t,x)$ in parallel using the same sequence of outcomes from a normally distributed random variable.

Often, a so-called *centred* approximation of the derivative is preferred, which in our context entails the expression

$$\frac{\partial C(t,x)}{\partial x} \approx \frac{C(t,x+h) - C(t,x-h)}{2h}.$$

A centred approximation gives more accurate estimates of the delta. We refer to [11] for a thorough analysis of numerical differentiation in finance and how

---

[4] In Exercise 5.3 we study how other derivatives of the claim can be simulated by this algorithm.

5.1.2 The Accuracy of Monte Carlo Methods

The accuracy of the Monte Carlo algorithm relies on the choice of N, the number of outcomes from the random variable Y. The larger N we choose, the closer we get to the true value that we want to approximate. But given an error tolerance, how large a value of N should we choose? Before discussing this question, let us note that both $C(t,x)$ and $\partial C(t,x)/\partial x$ can be written as

$$e^{-r(T-t)} \mathbb{E}_\mathcal{Q}[F(Y)],$$

for a suitable function F and a normally distributed random variable Y. Part 2 of Algorithm 1 can then be compactly written as

$$e^{-r(T-t)} \frac{1}{N} \sum_{k=1}^{N} F(y^k). \qquad (5.3)$$

Since the y^k's are independent outcomes from the same probability distribution, the central limit theorem (Thm. 1.1) implies that (5.3) is converging to a normal random variable with expectation $\mathbb{E}_\mathcal{Q}[F(Y)]$ and standard deviation

$$\varepsilon_N := \frac{\text{std}(F(Y))}{\sqrt{N}}.$$

Hence, the number we find in (5.3) will converge to the expectation we sought, but having random deviations of order $1/\sqrt{N}$ away from this number. These deviations become smaller when N gets larger. We say that the convergence rate of the Monte Carlo algorithm is of order $1/\sqrt{N}$. However, note that the error we calculate is stochastic. We can claim with only 68.3% probability that our approximation is within one standard deviation from the expectation we want to estimate. This means that, for each choice of N, we will obtain a number that is in the interval

$$(\mathbb{E}_\mathcal{Q}[F(Y)] - \varepsilon_N, \mathbb{E}_\mathcal{Q}[F(Y)] + \varepsilon_N),$$

with only 68.3% probability. The approximation will be within two standard deviations from the expectation with 95.4% probability. There is always a small chance that our estimate can be far from the value we want to find. However, this probability decreases with N and becomes very small when N is reasonably large. A convergence rate of order $1/\sqrt{N}$ is considered to be rather slow.

To control the error of the Monte Carlo algorithm exactly, we need to find the standard deviation of $F(Y)$. However, if we can calculate this, then

5.1 Pricing and Hedging with Monte Carlo Methods

we can in principle also calculate the expectation of $F(Y)$, which is the one we want to approximate numerically. In general we are not able to find the expectation of $F(Y)$, and henceforth we cannot find $\text{std}(F(Y))$. Using a numerical approximation of this leads us into a circle, since the error of this will depend on the standard deviation of $F^2(Y)$, and so on. Thus, in many practical situations we have only a qualitative understanding of the error in the Monte Carlo simuations. It is a drawback with the Monte Carlo algorithm that we cannot find an exact error bound as a function of N. However, the method has two clear advantages: firstly it is very simple to implement, and secondly we increase the accuracy by adding more outcomes from the normal random variable. We do not need to restart the whole simulation if we want better accuracy. Most other methods, including the finite difference method that we will consider next, must be completely restarted in order to increase the accuracy.

We end this subsection with an application of the Monte Carlo method on call options. Assume that we want to find the price of a call option with exercise in 100 days ($T = 100$) and strike price equal to 100 ($K = 100$). Furthermore, we assume for simplicity that the interest rate is $r = 0$ and put the value of the underlying stock equal to $S(0) = 100$. Finally, the daily volatility is put equal to $\sigma = 0.01$. The contract will have a payoff function $f(S(T)) = \max(0, S(T) - 100)$, or in terms of the risk-neutral Brownian motion $W(t)$

$$f(S(T)) = \max\left(0, 100\exp\left(-0.5(0.01)^2 T + 0.01 W(T)\right) - 100\right).$$

The call option price can be derived directly from the Black & Scholes formula, yielding the price

$$P_{\text{BS}}(0) = 100\left(\Phi(0.05) - \Phi(-0.05)\right) = 3.99.$$

Thus, it costs nearly €4 to have the right to buy the stock for €100 in 100 days. In Table 5.1 we list Monte Carlo simulated prices together with the error relative to the Black & Scholes price (that is, the difference between Black & Scholes and simulated price divided by the Black & Scholes price). The random nature of the Monte Carlo method is apparent from the stochastically fluctuating errors.

A small warning at the very end: if you try out the Monte Carlo algorithm on this example yourself, you will most likely get different numbers. Your computer will probably not give the same random numbers as the one used to obtain the figures in Table 5.1.

5.1.3 Pricing of Contingent Claims on Many Underlying Stocks

It is fairly straightforward to extend the Monte Carlo algorithm to options written on many underlying stocks. Recall the risk-neutral multi-dimensional stock price dynamics in (4.32)

Table 5.1. Monte Carlo simulated prices of a call option

N	MC price	Error,%
10^2	3.42	14.3
10^3	4.11	-3.01
10^4	4.07	-2.01
10^5	3.98	0.25
10^6	3.99	0.00

$$\mathrm{d}S_i(t) = rS_i(t)\,\mathrm{d}t + S_i(t)\sum_{j=1}^n \sigma_{ij}\,\mathrm{d}W_j(t),$$

for $i = 1,\ldots,n$, with $W_j(t)$ being independent Brownian motions with respect to $\mathcal{Q}$. From Sect. 4.7 we remember that the price of a claim with payoff function $X = f(S_1(T),\ldots,S_n(T))$ is given by

$$P(t) = C(t, S_1(t),\ldots,S_n(t)),$$

where

$$C(t,x_1,\ldots,x_n) = \mathrm{e}^{-r(T-t)}\mathbb{E}_{\mathcal{Q}}\left[f(S_1^{t,x_1}(T),\ldots,S_n^{t,x_n}(T))\right],$$

and

$$S_i^{t,x_i}(T) = x_i \exp\left(\left(r - \frac{1}{2}\sum_{j=1}^n \sigma_{ij}^2\right)(T-t) + \sum_{j=1}^n \sigma_{ij}(W_j(T) - W_j(t))\right).$$

We can simulate $C(t,x_1,\ldots,x_n)$ with a simple extension of the one-dimensional Monte Carlo algorithm:

Algorithm 2 *Monte Carlo algorithm for claims on many underlying stocks.*

1. *Draw nN independent outcomes from the random variable $Y \sim \mathcal{N}(0,1)$:*

$$(y_1^1,\ldots,y_n^1),\ldots,(y_1^N,\ldots,y_n^N).$$

2. *Calculate*

$$\mathrm{e}^{-r(T-t)}\frac{1}{N}\sum_{k=1}^N f(s_1^k, s_2^k,\ldots,s_n^k),$$

   where

$$s_i^k = x_i \exp\left(\left(r - \frac{1}{2}\sum_{j=1}^n \sigma_{ij}^2\right)(T-t) + \sqrt{T-t}\sum_{j=1}^n \sigma_{ij}\,y_j^k\right),$$

   for $i = 1,\ldots,n$.

The algorithm says that we have to draw nN independent outcomes from a standard normal distribution. This number is huge if we have many underlying stocks and want a reasonable accuracy for the price. The method becomes computationally demanding and inefficient, and techniques to speed up the simulations become necessary for practical applicability. One alternative is the so-called *quasi-Monte Carlo* method. The quasi-Monte Carlo method simulates the sum (5.3) by sampling the outcomes in a smart way.[5] We refer to [35], where this and other improvements of the Monte Carlo method are discussed. See also [14, 15], where adaptive methods are developed in conjuction with quasi-Monte Carlo techniques.

5.1.4 Pricing of Path-Dependent Claims

In this subsection we derive Monte Carlo algorithms for numerical pricing of two path-dependent claims called a knock-out and an Asian option. The typical feature of path-dependent claims is that we need to know the path of the underlying stock in order to find the payoff from the claim. This means that we need to know $S(t)$ for all times t between entry of the contract $t = 0$ up to exercise time $t = T$ to price the derivative contract.

A knock-out option is a claim that becomes worthless (i.e, is "knocked-out") when the price of the underlying stock breaks a barrier specified in the contract. We consider the special case of a call option which is knocked-out if the price of the underlying stock becomes greater than b during the life of the option. If $S(t) \leq b$ for all $t \in [0,T]$, the owner of the contract receives $\max(0, S(T) - K)$, where K is the agreed strike price. On the other hand, if $S(t) > b$ at least once, the option is cancelled and the holder receives nothing. Such a contract protects the issuer from paying out money if the stock price breaks the tolerance limit b. The payoff function from the knock-out option is

$$X = \max\left(0, S(T) - K\right) 1_{\{S(t) \leq b\,:\, t \in [0,T]\}},$$

where $1_{\{.\}}$ is the indicator function, being 1 when $\{S(t) \leq b : t \in [0,T]\}$ occurs, and zero otherwise. The contract only has sense if $b > K$, which we assume to be valid. From the option pricing theory in Chap. 4 it holds that

$$P(0) = e^{-rT} \mathbb{E}_Q \left[\max\left(0, S(T) - K\right) 1_{\{S(t) \leq b\,:\, t \in [0,T]\}} \right]. \quad (5.4)$$

We will now develop a Monte Carlo algorithm for finding $P(0)$. To simulate an outcome from the payoff X we must first check if $S(t)$ has violated the barrier b for any $t \in [0,T]$. If this is not the case, we simulate $S(T)$ in order to find the payoff. We will solve this by simulating $S(t)$ recursively.

Partition the interval $[0,T]$ in $M := T/\theta$ non-overlapping subintervals, where θ is a small number. Set $t_j = \theta j$, for $j = 0, 1, \ldots, M$. Since $S(t)$ is

---

[5] In fact, the outcomes are drawn according to a deterministic scheme called a *low discrepancy sequence*.

$$S(t) = S(0) \exp\left(\left(r - \frac{1}{2}\sigma^2\right)t + \sigma W(t)\right),$$

we find for $j > 0$,

$$\begin{aligned}
S(t_j) &= S(0) \exp\left(\left(r - \frac{1}{2}\sigma^2\right) t_j + \sigma W(t_j)\right) \\
&= S(0) \exp\left(\left(r - \frac{1}{2}\sigma^2\right)(t_{j-1} + \theta) + \sigma(W(t_{j-1}) + W(t_j) - W(t_{j-1}))\right) \\
&= S(0) \exp\left(\left(r - \frac{1}{2}\sigma^2\right) t_{j-1} + \sigma W(t_{j-1})\right) \\
&\quad \times \exp\left(\left(r - \frac{1}{2}\sigma^2\right)\theta + \sigma(W(t_j) - W(t_{j-1}))\right) \\
&= S(t_{j-1}) \exp\left(\left(r - \frac{1}{2}\sigma^2\right)\theta + \sigma\sqrt{\theta}\, Y_j\right),
\end{aligned}$$

for $j = 1, \ldots, M$. The random variables $Y_j = \theta^{-1/2}(W(t_j) - W(t_{j-1}))$ are distributed according to a standard normal distribution and independent of $S(t_{j-1})$. This recursion formula paves the way for a simple Monte Carlo approach to simulate the path of $S(t)$ at the times $t_0, t_1, t_2, \ldots, t_M$, and thereby provides information about whether the path has broken the barrier b or not. Of course $S(t)$ can become greater than b between two time instants t_j and t_{j+1}, and decrease sufficiently in order to become less than b before t_{j+1} is reached. If this happens we will not detect the breaking of the barrier because we only sample the path at the discrete time instants t_j and t_{j+1}. However, the probability that this will happen is very small when we sample densely in time. Hence, the error introduced from discrete sampling of the path is small when approximating the expectation in (5.4) using the recursive formula for $S(t_j)$.

We simulate an outcome from the random variable (the payoff) X by simulating the path $S(t)$ at the discrete time points t_j, $j = 0, 1, \ldots, M$, and checking if the barrier is violated or not. In the latter case, the outcome is $\max(0, S(T) - K)$, and zero otherwise. Hence, the algorithm becomes:

Algorithm 3 *Monte Carlo algorithm for a knock-out option.*

1. For $k = 1, \ldots N$,
    a. For $j = 1, \ldots M$,
        – Draw one outcome y_j^k from the random variable $Y_j \sim \mathcal{N}(0, 1)$.
        – Calculate
        $$s_j^k = s_{j-1}^k \exp\left(\left(r - \frac{1}{2}\sigma^2\right)\theta + \sigma\sqrt{\theta}\, y_j^k\right).$$
        – If $s_j^k > b$, let $x^k = 0$ and return to 1.

b. Let $x^k = \max(0, s_M^k - K)$.

2. Calculate $e^{-rT} \frac{1}{N} \sum_{k=1}^{N} x^k$.

We emphasize that step 1b is only reached if all the s_j^k's, for $j = 1, \ldots, M$, pass the test in step 1a.

Below we suggest a Visual Basic implementation of the algorithm. Since $\max(a, b)$ is not a predefined function in Visual Basic, the reader must define this. We assume it is implemented as a function Max() below:

```
Function KnockOutCallMC(S0, r, sigma, Ks, et, b, N, M)
    ' S0=current stock price
    ' r=interest rate
    ' sigma=volatility
    ' Ks=strike price
    ' et=time of exercise
    ' b=barrier
    ' N=number of Monte Carlo replications
    ' M=partition of time interval

    p=0
    For k=1 To N
      s=S0
      j=0
      While (s<b And j<M)
        j=j+1
        u=Rnd()
        y=Application.WorksheetFunction.NormSInv(u)
        s=s*Exp(((r-0.5*sigma*sigma)*et/M)+
            sigma*Sqr(et/M)*y)
      Wend
      If (j=M And s<b) Then
        p=p+Max(0,s-K)
      End If
    Next
    p=p/N
    KnockOutCallMC=Exp(-r*et)*p
End Function
```

The variable s keeps track of the stock price when it runs over different time instances j. In the "While"-loop we simulate the path and at the same time check whether the stock crosses the barrier or not. If it survives through the loop, it means that the path is below b except possibly at the end point

$j = M$. In the "if"-test after the loop we check if the option is "knocked-out" at the time of exercise before calculating the outcome of the payoff.

We continue by developing a Monte Carlo algorithm for the price of an Asian option, also known as an average option. There exist many different types of Asian options, but we are going to concentrate on a call option written on the average of the stock price from zero to time of exercise. Mathematically, we write the payoff function from the Asian option of interest as

$$X = \max\left(0, \frac{1}{T}\int_0^T S(t)\,\mathrm{d}t - K\right). \tag{5.5}$$

In practice we cannot integrate the stock price over $[0, T]$, but measure it instead at contracted discrete time points and find the average over these.[6] In other words, we consider instead a discretized version of (5.5),

$$X = \max\left(0, \frac{1}{M+1}\sum_{j=0}^M S(t_j) - K\right), \tag{5.6}$$

where the $M+1$ times for measurement of the stock price are specified in the option contract. The initial time $t = 0$ is not always taken into account. However, the considerations below are easily modified for this case. In contrast with the knock-out option, we do not necessarily have a uniform partition of the time interval $[0, T]$. The time partition is assumed to be increasing in the sense that $0 = t_0 < t_1 < \ldots < t_{M-1} < t_M = T$. We remark that (5.6) is often considered as an approximation of (5.5), even though practitioners are mostly concerned with average options on the form (5.6) and (5.5) only seems to be of mathematical interest.

We now create a Monte Carlo algorithm that approximates the price $P(0)$ of the Asian option with payoff (5.6),

$$P(0) = e^{-rT}\mathbb{E}_{\mathcal{Q}}\left[\max\left(0, \frac{1}{M+1}\sum_{k=0}^M S(t_j) - K\right)\right].$$

Let $\theta_j := t_j - t_{j-1}$. Similar to above, we find

$$S(t_j) = S(t_{j-1})\exp\left(\left(r - \frac{1}{2}\sigma^2\right)\theta_j + \sigma\sqrt{\theta_j}Y_j\right),$$

where the random variables $Y_j = \theta_j^{-1/2}(W(t_j) - W(t_{j-1}))$ are distributed according to a standard normal distribution and independent of $S(t_{j-1})$. We now make a recursion formula which calculates the sum of stock prices. To motivate this recursion, assume for a moment that $M = 3$: it is easily seen that with the notation $\tilde{Y}_j := (r - \sigma^2/2)\theta_j + \sigma\sqrt{\theta_j}Y_j$, we get

---

[6] Options where the average is taken over a discrete number of prices are known as *Bermudan* options.

$$\frac{1}{4}\sum_{j=0}^{3} S(t_j) = \frac{1}{4}\left(S(t_0) + S(t_1) + S(t_2) + S(t_3)\right)$$
$$= \frac{S(0)}{4}\left(1 + \exp(\tilde{Y}_1) + \exp(\tilde{Y}_1 + \tilde{Y}_2) + \exp(\tilde{Y}_1 + \tilde{Y}_2 + \tilde{Y}_3)\right)$$
$$= \frac{S(0)}{4}\left(1 + \exp(\tilde{Y}_1)\left(1 + \exp(\tilde{Y}_2)\left(1 + \exp(\tilde{Y}_3)\right)\right)\right).$$

This motivates the following representation of the sum $\frac{1}{M+1}\sum_{j=0}^{M} S(t_j)$,

$$\frac{1}{M+1}\sum_{j=0}^{M} S(t_j) = \frac{S(0)}{M+1}Z_1,$$

where the random variable Z_1 is defined from the (backward) recursion

$$Z_M = 1 + \exp\left(\tilde{Y}_M\right), \tag{5.7}$$
$$Z_{j-1} = 1 + \exp\left(\tilde{Y}_{j-1}\right)Z_j, \quad j = M, M-1, \ldots, 2. \tag{5.8}$$

The payoff function in (5.6) can thus be written

$$X = \max\left(0, \frac{S(0)}{M+1}Z_1 - K\right).$$

Then the Monte Carlo algorithm becomes:

Algorithm 4 *Monte Carlo algorithm for an Asian option.*

1. For $k = 1, \ldots, N$
   a. Draw M independent outcomes from the random variable $Y \sim \mathcal{N}(0,1)$:
   $$(y_1^k, \ldots, y_M^k).$$
   b. Set, for $j = 1, \ldots, M$
   $$\tilde{y}_j^k = \left(r - \frac{1}{2}\sigma^2\right)\theta_j + \sigma\sqrt{\theta_j}y_j^k.$$
   c. Set $z_M^k = 1 + \exp\left(\tilde{y}_M^k\right)$.
   d. For $j = M, \ldots, 2$,
      - Calculate $z_{j-1}^k = 1 + \exp\left(\tilde{y}_{j-1}^k\right) z_j^k$.
   e. Set $x^k = \max\left(0, \frac{S(0)}{M+1} z_1^k - K\right)$.
2. Calculate $e^{-rT} \frac{1}{N}\sum_{k=1}^{N} x^k$.

112   5 Numerical Pricing and Hedging of Contingent Claims

We suggest an implementation of this algorithm in Visual Basic. In the code we assume that the time interval $[0, T]$ is uniformly partitioned, that is, $t_j = \theta j$ where $\theta = T/M$ for an M chosen by the user.

```
Function AsianCallMC(S0, r, sigma, Ks, et, N, M)
    ' S0=current stock price
    ' r=interest rate
    ' sigma=volatility
    ' Ks=strike price
    ' et=time to exercise
    ' N=number of Monte Carlo replications
    ' M=partition of time interval

p=0
For k=1 To N
   For j=1 To M
      u=Rnd()
      y=Application.WorksheetFunction.NormSInv(u)
      ytilde=((r-0.5*sigma*sigma)*et/M)+
         sigma*Sqr(et/M)*y
      z=1+Exp(ytilde)*z
   Next
   p=p+Max(0,S0*z/(M+1)-K)
Next
p=p/N
AsianCallMC=Exp(-r*et)*p
End Function
```

Notice that the inner "for"-loop running over j's is not counting backwards as in the algorithm. In the implementation we can let it go forward, while we let the algorithm follow the mathematical derivations closely.

In this subsection we have only considered Monte Carlo routines for approximating the *price* of two path-dependent options. Those interested in methods to calculate the hedging strategy are referred to [27]. There the authors use the Malliavin derivative encountered in Sect. 4.5 to derive Monte Carlo algorithms for calculating the delta for various path-dependent options, for instance, Asian options.

5.2 Pricing and Hedging with the Finite Difference Method

Recall the Black & Scholes partial differential equation (4.12),

5.2 Pricing and Hedging with the Finite Difference Method

$$\frac{\partial C(t,x)}{\partial t} + rx\frac{\partial C(t,x)}{\partial x} + \frac{1}{2}\sigma^2 x^2 \frac{\partial^2 C(t,x)}{\partial x^2} = rC(t,x),$$

where $x \geq 0$, $t \in [0,T]$ and $C(T,x) = f(x)$. The solution $C(t,x)$ gives the price of an option with payoff $f(S(T))$ at exercise time T, when the underlying stock has value x at the current time t. An alternative to the Monte Carlo approach is to solve this partial differential equation numerically to find the price (and hedge ratio) of the option. The finite difference method starts out by approximating the derivatives of $C(t,x)$ by so-called *finite differences*, which in combination with the Black & Scholes partial differential equation lead to a recursive scheme for the price. The finite difference method is computationally far more effective than the Monte Carlo technique, but requires more preparation. There exist many variants of the method, but we restrict our attention to the simplest one.

Before introducing the finite difference technique, it is advantageous to simplify the Black & Scholes partial differential equation by a change of variables. Let

$$u(\tau, y) := \exp(ay + b\tau) C\left(T - \frac{2\tau}{\sigma^2}, e^y\right), \tag{5.9}$$

with

$$a = \frac{1}{2}\left(\frac{2r}{\sigma^2} - 1\right), \text{ and } b = \frac{1}{4}\left(\frac{2r}{\sigma^2} + 1\right)^2.$$

The variable y runs over the real numbers $\mathbb{R}$, while $\tau \in [0, \sigma^2 T/2]$. It turns out that $u(\tau, y)$ is the solution of the so-called heat equation[7] which we now show. The chain rule yields the derivatives (see Exercise 5.2),

$$\frac{\partial u(\tau,y)}{\partial \tau} = bu(\tau,y) - \frac{2}{\sigma^2}\exp(ay+b\tau)\frac{\partial C(\tau,y)}{\partial t},$$

$$\frac{\partial u(\tau,y)}{\partial y} = au(\tau,y) + \exp((a+1)y+b\tau)\frac{\partial C(\tau,y)}{\partial x},$$

$$\frac{\partial^2 u(\tau,y)}{\partial y^2} = (2a+1)\frac{\partial u(\tau,y)}{\partial y} - (a^2+a)u + \exp((a+2)y+b\tau)\frac{\partial^2 C(\tau,y)}{\partial x^2}.$$

Since $C(t,x)$ solves Black & Scholes' partial differential equation, we find that (see Exercise 5.2)

$$\frac{\partial u(\tau,y)}{\partial \tau} = \frac{\partial^2 u(\tau,y)}{\partial y^2}. \tag{5.10}$$

Further, u satisfies the *initial* condition

$$u(0,y) = e^{ay} f(e^y).$$

If we know the solution of (5.10), we can calculate the price $C(t,x)$ from the formula

---

[7] Recall the discussion in Subsect. 4.3.2.

$$C(t,x) = \exp\left(-a\ln x - \frac{1}{2}b\sigma^2(T-t)\right) \times u\left(\frac{1}{2}\sigma^2(T-t), \ln x\right). \quad (5.11)$$

Letting $x = S(0)$ and $t = 0$, we find the price of the option as

$$P(0) = C(0, S(0)) = \exp\left(-a\ln S(0) - \frac{1}{2}b\sigma^2 T\right) \times u\left(\frac{1}{2}\sigma^2 T, \ln S(0)\right). \quad (5.12)$$

Instead of solving numerically the Black & Scholes partial differential equation, we consider the simpler equation (5.10).

We partition the (y, τ)-domain into a regular grid (see Fig. 5.1). In each point on the grid (that is, where two grid lines intersect) we use (5.10) to derive a recursion from which we can derive approximate values of u. We choose θ to be the distance between each point along the τ-axis, while h is the distance in the y-direction. The accuracy of our approximation will of course depend on how small we choose θ and h to be. We set $\tau_k := k \cdot \theta$ and $y_i = i \cdot h$, and denote the approximate value of $u(\tau_k, y_i)$ by u_i^k. The natural number k ranges from 0 up to M, where $\tau_M = \sigma^2 T/2$. Since the partial differential equation for u is defined for all real y, we should in principle find an approximation of (τ_k, y_i) for all y_i with i ranging over all integers. However, on a computer we must restrict ourselves to a finite amount of y_i's, and we therefore let i range over the integers from $-N$ to N, where $y_{-N} = -Nh = -y_{\max}$, $y_N = Nh = y_{\max}$ and $y_{\max}$ is some fixed limit for the y-values of interest. If we, for instance, are concerned with calculating $C(0, S(0))$, the limit $y_{\max}$ should be chosen significantly greater than $|\ln S_0|$ in order to have a good approximation of u when calculating C from (5.12).

We approximate the derivatives of u with so-called finite differences:

$$\frac{\partial u(\tau_k, y_i)}{\partial \tau} \approx \frac{u(\tau_k + \theta, y_i) - u(\tau_k, y_i)}{\theta} \approx \frac{u_i^{k+1} - u_i^k}{\theta},$$

$$\frac{\partial u(\tau_k, y_i)}{\partial x} \approx \frac{u(\tau_k, y_i + h) - u(\tau_k, y_i - h)}{2h} \approx \frac{u_{i+1}^k - u_{i-1}^k}{2h},$$

$$\frac{\partial^2 u(\tau_k, y_i)}{\partial x^2} \approx \frac{u(\tau_k, y_i + h) - 2u(\tau_k, y_i) + u(\tau_k, y_i - h)}{h^2}$$

$$\approx \frac{u_{i+1}^k - 2u_i^k + u_{i-1}^k}{h^2}.$$

Inserting these expressions into (5.10), we obtain the following equation,

$$\frac{u_i^{k+1} - u_i^k}{\theta} = \frac{u_{i+1}^k - 2u_i^k + u_{i-1}^k}{h^2},$$

or, in a recursive form,

$$u_i^{k+1} = \frac{\theta}{h^2}u_{i+1}^k + \left(1 - \frac{2\theta}{h^2}\right)u_i^k + \frac{\theta}{h^2}u_{i-1}^k. \quad (5.13)$$

5.2 Pricing and Hedging with the Finite Difference Method

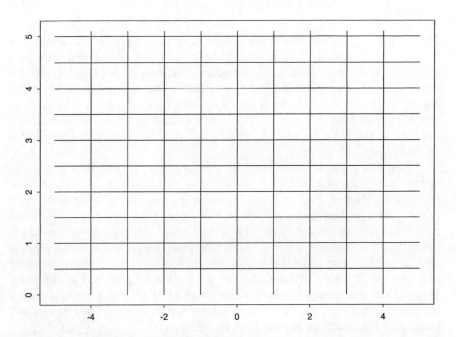

Fig. 5.1. An example of a grid where the horizontal y-axis is partitioned with distance 1 and the vertical τ-axis with distance 1/2

To start the recursion, we need to specify the u_i^0's. The value of u at time zero provides us with these numbers: we know that $u(0,y) = \exp(ay)C(T, e^y) = \exp(ay)f(e^y)$, and therefore it follows that

$$u_i^0 = e^{a \cdot ih} f(e^{ih}),$$

for $i = -N+1, \ldots, N-1$. In addition, we have to specify u_i^k along the boundary of the grid, that is, for $k = 1, \ldots, M$ when $i = -N$ or $i = N$. We find these values from asymptotic considerations of the price function $C(t,x)$. First, consider the limit of $C(T - \sigma^2\tau/2, e^y)$, when $y \to -\infty$:

$$\lim_{y \to -\infty} C\left(T - \frac{1}{2}\sigma^2\tau, e^y\right) = \exp\left(-r\left(T - \left(T - \frac{1}{2}\sigma^2\tau\right)\right)\right)$$
$$\times \mathbb{E}_Q\left[f(S^{T-\sigma^2\tau/2,0}(T))\right]$$
$$= f(0)\exp\left(-\frac{1}{2}r\sigma^2\tau\right).$$

A natural boundary condition on the left-hand side of the grid is therefore

$$u^k_{-N} = f(0) \exp\left(\left(b - \frac{1}{2}r\sigma^2\right)k\theta - aNh\right), \qquad (5.14)$$

for $k = 1, 2, ..., M$. Note that when $f(0) = 0$, we have $C(T - \sigma^2\tau/2, \exp(y)) \to 0$, when $y \to -\infty$. If at the same time $a < 0$, we cannot conclude in a straightforward manner that the boundary condition should be $u^k_{-N} = 0$, for $k = 1, 2, ..., M$. The condition (5.14) is based on the assumption $a > 0$ in the case of $f(0) = 0$ since the term $\exp(-ay)$ will tend to infinity when $a < 0$. If it turns out that $a < 0$ and $f(0) = 0$, we must go into a more thorough analysis of the limit for $u(\tau, y)$, when $y \to -\infty$. We shall not address this problem here.

Along the right boundary of the grid we need information about the asymptotic behaviour of $C(t, x)$, when $x \to \infty$, which clearly depends on the payoff function f. For example, when we consider a call option with payoff function $f(x) = \max(0, x - K)$, it holds that $C(t, x) \sim x - K$ for $x \to \infty$. This can be seen from Black & Scholes' formula, or we can argue heuristically as follows: when $x \to \infty$, the underlying stock is worth a great deal more than the strike price and the option will almost with certainty be exercised at time T. Hence, the fair price must be close to the difference between the underlying stock and the strike price K. For a general option we *assume* that $C(t, x) \sim c(x)$ for a known function $c(x)$, when x becomes large. The boundary values can thus be written

$$u^k_N = \exp(aNh + bk\theta)c(e^{Nh}), \qquad (5.15)$$

for $k = 1, 2, ..., M$. We are now ready to present the finite difference algorithm for finding approximately the values of $u(\tau, y)$ in certain grid points.

Algorithm 5 *Algorithm for the finite difference method.*

1. Choose M and N.
2. For $k = 1, 2, ..., M$, set

$$u^k_{-N} = f(0) \exp\left(\left(b - \frac{1}{2}r\sigma^2\right)k\theta - aNh\right),$$
$$u^k_N = \exp\left(aNh + bk\theta\right)c(e^{Nh}).$$

3. For $i = -N, ..., N$, set

$$u^0_i = \exp\left(aih\right) f(\exp(ih)).$$

4. For $k = 1, ..., M - 1$, $i = -N + 1, N - 1$,

$$u^{k+1}_i = \frac{\theta}{h^2} u^k_{i+1} + \left(1 - \frac{2\theta}{h^2}\right) u^k_i + \frac{\theta}{h^2} u^k_{i-1}.$$

5.2 Pricing and Hedging with the Finite Difference Method

This algorithm provides us with values for u_i^k for all grid points. We find an approximation of $C(t,x)$ via (5.11) in the following manner: for a given coordinate (t,x), search for the coordinate (τ_k, y_i) in the grid which is closest to $(\sigma^2(T-t)/2, \ln x)$. Then, approximate $C(t,x)$ with the value

$$C(t,x) \approx \exp(-ay_i - b\tau_k) u(\tau_k, y_i),$$

invoking this choice of τ_k and y_i.

We discuss an important issue related to numerical stability. An unlucky choice of the discretization parameters θ and h can lead to rapidly fluctuating values of u_i^k which have nothing to do with the actual solution. To ensure that our algorithm has a stable behaviour, we must choose θ and h such that the sign of the constant in front of u_i^k on the right-hand side (5.13) is positive. That is, we must choose θ and h such that

$$\theta \leq h^2/2. \tag{5.16}$$

We note that if we make h smaller in order to increase the accuracy of our calculations, we have to make θ "quadratically" smaller in order to avoid numerical instability.

If we want to find the replicating strategy, we have to calculate $\partial C(t,x)/\partial x$. From (5.11) it holds that

$$\frac{\partial C(t,x)}{\partial x} = x^{-1} \exp\left(-a \ln x - \frac{1}{2} b\sigma^2(T-t)\right) \times \frac{\partial u(\sigma^2(T-t)/2, \ln x)}{\partial y}$$
$$- ax^{-1} C(t,x).$$

Based on centred differences, a numerical approximation of $\partial u(\tau, y)/\partial y$ is

$$\frac{\partial u(\tau_k, y_i)}{\partial y} \approx \frac{u_{i+1}^k - u_{i-1}^k}{2h},$$

which we can use when calculating $\partial C(t,x)/\partial x$.

Below we suggest an implementation of the finite difference method for call options. The implementation is written as a Sub() in Visual Basic, and outputs the result as a matrix in the Excel spreadsheet. When we consider call options, we know that the function c in the algorithm is $c(x) = x - K$, where K is the strike price.

```
Sub CallFD(r, sigma, Ks, et, N, M, ymax)
    ' r=interest rate
    ' sigma=volatility
    ' Ks=strike price
    ' et=time of exercise
    ' N=grid in y
    ' M=grid in τ
```

```
' ymax=maximal length of grid in y

Dim u(100,500) As Double
Dim C(100,500) As Double
a=0.5*(2*r/(sigma*sigma)-1)
b=0.25*(2*r/(sigma*sigma)+1)^2
theta=0.5*sigma*sigma*et/M
h=0.5*ymax/N
For k =0 To M
   u(k,0)=0
   u(k,2*N)=Exp(a*N*h+b*k*theta)*(Exp(N*h)-Ks)
Next
For i=1 To 2*N-1
   u(0,i)=Exp(a*(i-N)*h)*Max(0,Exp(h*(i-N))-Ks)
Next
d=theta/(h*h)
For k=0 To M-1
   For i=1 To 2*N-1
      u(k+1,i)=d*u(k,i+1)+(1-2*d)*u(k,i)+d*u(k,i-1)
   Next
Next
'Transforming back to C
For k=0 To M
   For i=0 To 2*N
      C(k,i)=Exp(-a*h*(i-N)-b*theta*(M-k))*u(M-k,i)
   Next
Next
'Write the results onto ``Sheet1'' in the spreadsheet
For i=-N To N
   x=Exp(h*i)
   Worksheets(``Sheets1'').Cells(M+3,N+i+2).Value=x
Next
For k=0 To M-1
   t=eT-(theta*k)/(0.5*sigma*sigma)
   Worksheets(``Sheets1'').Cells(k+2,1).Value=t
   For i=0 To 2*N
      Worksheets(``Sheets1'').Cells(k+2,N+i+2).Value=
         C(M-k,i)
   Next
Next
End Sub
```

We assume that the matrices u and C are appropriately sized in Visual Basic by the user. If we use CallFD from a Main()-routine, option prices for different times and values of the underlying stock are written into the spreadsheet.

5.2 Pricing and Hedging with the Finite Difference Method

The finite difference method can also be applied to price options written on many underlying stocks. However, it becomes very inefficient when the dimension becomes 4 or higher. Worth noticing is that the Monte Carlo algorithm is not dependent on the number of underlying stocks. On the other hand, the finite difference method is extremely efficient and provides us with a range of prices rather than just one price for a given coordinate (t, x).

Exercises

5.1 Let U be a uniformly distributed random variable on the interval $[0, 1]$. Show that $Y = \Phi^{-1}(U)$ is a standard normal distributed random variable, where $\Phi(x)$ is defined in Sect. 1.2.

5.2 Calculate the derivatives $\partial u/\partial \tau$, $\partial u/\partial y$ and $\partial^2 u/\partial y^2$ for u defined in (5.9). Show that $\partial u(\tau, y)/\partial \tau = \partial^2 u(\tau, y)/\partial y^2$.

5.3 Use the density approach in Subsect. 4.3.5 to derive expressions for the Greeks[8] vega and rho of an option on a stock with payoff $f(S(T))$:

$$\text{vega} = \frac{\partial C(t, x)}{\partial \sigma}, \quad \text{rho} = \frac{\partial C(t, x)}{\partial r}.$$

Explain how you can simulate these using the Monte Carlo method.

5.4 Implement Algorithm 1 in Visual Basic (or any other software). Use this function to price a so-called *quadratic option* which has payoff

$$f(x) = \{\max(0, x - K)\}^2,$$

that is, the square of a call option. To obtain a reasonable accuracy, first find N for a call option that gives a satisfactory accuracy of the price. Use the same N to price the quadratic option.

5.5 Make an algorithm that simulates a *path* of Brownian motion and geometric Brownian motion.

---

[8] It is a curiosity that "vega" is *not* a letter in the Greek alphabet, but seems to be invented as such in finance.

A Solutions to Selected Exercises

1.1

From the definition of expectation and the probability density of X we have

$$\mathbb{E}[X] = \frac{1}{\sqrt{2\pi\sigma^2}} \int_{-\infty}^{\infty} x \exp\left(-\frac{1}{2\sigma^2}(x-\mu)^2\right) dx$$

$$= \frac{1}{\sqrt{2\pi\sigma^2}} \int_{-\infty}^{\infty} z \exp\left(-\frac{1}{2\sigma^2}z^2\right) dx + \mu \frac{1}{\sqrt{2\pi\sigma^2}} \int_{-\infty}^{\infty} \exp\left(-\frac{1}{2\sigma^2}z^2\right) dx$$

$$= \mu,$$

where we used the change of variables $z = x - \mu$. The variance is calculated similarly.

1.2

Use the linearity of the expectation to find

$$\mathbb{E}\left[(X - \mathbb{E}[X])^2\right] = \mathbb{E}[X^2] - 2\mathbb{E}[X\mathbb{E}[X]] + \left(\mathbb{E}[X]^2\right)^2$$
$$= \mathbb{E}[X^2] - \mathbb{E}[X]^2.$$

An alternative way to define covariance is

$$\text{Cov}(X, Y) = \mathbb{E}[(X - \mathbb{E}[X])(Y - \mathbb{E}[Y])].$$

1.3

Define $Y = a + bX$. We calculate the probability distribution of Y:

$$P(Y \leq y) = P(a + bX \leq y) = P(X \leq (y-a)/b) = \int_{-\infty}^{(y-a)/b} p_X(x)\, dx,$$

where p_X is the density function of X. Differentiating with respect to y yields

$$p_Y(y) = \frac{1}{b} p_X((y-a)/b) = \frac{1}{\sqrt{2\pi b^2 \sigma^2}} \exp\left(-\frac{(y-a-b\mu)^2}{2\sigma^2 b^2}\right),$$

which is the density function of a normally distributed random variable with mean $a + b\mu$ and variance $b^2\sigma^2$.

1.4

Define $Y = \exp(X)$, where $X \sim \mathcal{N}(\mu, \sigma^2)$. Then by definition Y is a lognormal variable. We find its probability density by first deriving the cumulative distribution function. Letting $y > 0$,

$$\begin{aligned}P_Y(y) &= \mathcal{P}\left(\exp(X) \leq y\right) \\ &= \mathcal{P}\left(X \leq \ln(y)\right) \\ &= \frac{1}{\sqrt{2\pi\sigma^2}} \int_{-\infty}^{\ln(y)} \exp\left(-\frac{1}{2\sigma^2}(x-\mu)^2\right) \, dx.\end{aligned}$$

Differentiating this with respect to y yields

$$p_Y(y) = \frac{1}{y\sqrt{2\pi\sigma^2}} \exp\left(-\frac{1}{2\sigma^2}(\ln(y)-\mu)^2\right).$$

1.5

Let X and Y have density functions $p_X(x)$ and $p_Y(y)$, resp. From the definition of the expectation we have

$$\begin{aligned}\mathbb{E}[XY] &= \int_{\mathbb{R}^2} xy p_{X,Y}(x,y) \, dxdy \\ &= \int_{\mathbb{R}^2} xy p_X(x) p_Y(y) \, dxdy \\ &= \int_{\mathbb{R}} x p_X(x) \, dx \int_{\mathbb{R}} y p_Y(y) \, dy \\ &= \mathbb{E}[X]\mathbb{E}[Y],\end{aligned}$$

where in the second equality we used that X and Y are independent.

1.6

We have that $\theta = (\mu, \sigma^2)'$, and the likelihood function is

$$\begin{aligned}L(\mathbf{x}; \mu, \sigma^2) &= \prod_{i=1}^{n} \frac{1}{\sqrt{2\pi\sigma^2}} \exp\left(-\frac{1}{2\sigma^2}(x_i - \mu)^2\right) \\ &= (2\pi\sigma^2)^{-n/2} \exp\left(-\frac{1}{2\sigma^2} \sum_{i=1}^{n}(x_i - \mu)^2\right).\end{aligned}$$

We optimize with respect to μ and σ^2 by differentiating and putting the derivative equal to zero:

$$\frac{\partial L(\mathbf{x}; \mu, \sigma^2)}{\partial \mu} = L(\mathbf{x}; \mu, \sigma^2)\left(-\frac{1}{\sigma^2} \sum_{i=1}^{n}(x_i - \mu)\right) = 0,$$

which yields
$$\widehat{\mu} = \frac{1}{n}\sum_{i=1}^{n} x_i.$$

We perform a similar argument for σ^2:
$$\frac{\partial L(\mathbf{x};\mu,\sigma^2)}{\partial \sigma^2} = L(\mathbf{x};\mu,\sigma^2)\left(-\frac{n}{2\sigma^2} + \frac{1}{2(\sigma^2)^2}\sum_{i=1}^{n}(x_i-\mu)^2\right) = 0,$$

which gives
$$\widehat{\sigma^2} = \frac{1}{n}\sum_{i=1}^{n}(x_i - \widehat{\mu})^2.$$

2.1

a) Since $B(0) = 0$, we find $B(t-s) = B(t-s) - B(0)$, which is a Brownian increment over the time interval $t - s - 0 = t - s$. By definition, $B(t-s)$ is normally distributed with expectation 0 and variance $t - s$, which is the same as the increment $B(t) - B(s)$.

b) Brownian increments are independent, and thus $B(t) - B(s)$ is independent of $B(s) = B(s) - B(0)$.

2.2

We calculate the nth moment of $S(t)$, $n \geq 1$:
$$\begin{aligned}
\mathbb{E}\left[S(t)^n\right] &= S(0)^n e^{\mu n t}\mathbb{E}\left[e^{\sigma n B(t)}\right] \\
&= S(0)^n e^{\mu n t}\frac{1}{\sqrt{2\pi t}}\int_{-\infty}^{\infty} e^{\sigma n x}e^{-x^2/2t}\,dx \\
&= S(0)^n e^{\mu n t}\frac{1}{\sqrt{2\pi t}}\int_{-\infty}^{\infty} \exp\left(-\frac{1}{2t}(x-\sigma n t)^2\right)\,dx\, e^{\frac{1}{2}\sigma^2 n^2 t^2} \\
&= S(0)^n e^{(\mu n + \frac{1}{2}\sigma^2 n^2)t}.
\end{aligned}$$

2.4

To simplify notation, let $q := \mathrm{VaR}_\alpha(t)$. We calculate
$$\begin{aligned}
1 - \alpha &= \mathcal{P}\left(S(t) \leq q\right) \\
&= \mathcal{P}\left(B(t) \leq \frac{1}{\sigma}\left(\ln\frac{q}{S(0)} - \mu t\right)\right).
\end{aligned}$$

But since $B(t) \sim \mathcal{N}(0,t)$, we can write $B(t) \stackrel{d}{=} \sqrt{t}X$ for $X \sim \mathcal{N}(0,1)$. Hence,

$$P\left(B(t) \le \frac{1}{\sigma}\left(\ln\frac{q}{S(0)} - \mu t\right)\right) = P\left(X \le \frac{1}{\sigma\sqrt{t}}\left(\ln\frac{q}{S(0)} - \mu t\right)\right) = 1-\alpha.$$

The $1-\alpha$ quantile of a standard normal distributed variable is denoted by q_α in the exercise. Therefore,

$$q_\alpha = \frac{1}{\sigma\sqrt{t}}\left(\ln\frac{q}{S(0)} - \mu t\right).$$

After reorganizing, we arrive at the desired expression.

2.5

We use Taylor's Formula of degree 1 with remainder on the function $\ln(1+z)$:

$$\ln(1+z) = z - \frac{1}{2(1+\tilde{z})^2}z^2,$$

where $|\tilde{z}| \le |z|$. The remainder is bounded in absolute value by $z^2/2$ when z is positive, and $z^2/2(1+z)^2$ when z is negative. In any case, when z is small, the remainder term becomes insignificant compared with z, and we have

$$\ln(1+z) \approx z,$$

to a high degree of accuracy.

Assume that we have observed stock prices $s(0), s(1), ..., s(N)$. Let $z = \frac{s(k)-s(k-1)}{s(k-1)}$, the return from the stock from time $k-1$ to time k. Reorganizing, we find $1+z = s(k)/s(k-1)$, and if the returns are small, it follows that

$$\ln\left(\frac{s(k)}{s(k-1)}\right) \approx \frac{s(k)-s(k-1)}{s(k-1)}.$$

Hence, the logreturns are approximately equal to the returns as long as the latter are small. How close they are can be evaluated by calculating the remainder.

2.6

From the definition of Brownian motion we know that $B(t+s) - B(t) \sim \mathcal{N}(0,s)$. For simplicity, define $X := B(t+s) - B(t)$. We calculate the kth moment of X using a table of integrals:

$$\begin{aligned}\mathbb{E}\left[X^k\right] &= \frac{1}{\sqrt{2\pi s}}\int_{-\infty}^{\infty} x^k e^{-x^2/2s}\,dx \\ &= \frac{1}{\sqrt{2\pi s}}(2s)^{(k+1)/2}\Gamma\left(\frac{k+1}{2}\right) \\ &= \frac{1}{\sqrt{\pi}}(2s)^{k/2}\Gamma\left(\frac{k+1}{2}\right),\end{aligned}$$

where Γ is the Gamma function.

3.1

The expression $\{\sum_{i=1}^{n} X(s_i)(B(s_{i+1}) - B(s_i))\}^2$ can be written as

$$\left\{\sum_{i=1}^{n} X(s_i)(B(s_{i+1}) - B(s_i))\right\}^2$$

$$= \sum_{i,j=1}^{n} X(s_i)X(s_j)(B(s_{i+1}) - B(s_i))(B(s_{j+1}) - B(s_j))$$

$$= \sum_{i=1}^{n} X^2(s_i)(B(s_{i+1}) - B(s_i))^2 + \sum_{i \neq j} X(s_i)X(s_j)$$
$$\times (B(s_{i+1}) - B(s_i))(B(s_{j+1}) - B(s_j))$$

$$= \sum_{i=1}^{n} X^2(s_i)(B(s_{i+1}) - B(s_i))^2 + 2\sum_{i<j} X(s_i)X(s_j)$$
$$\times (B(s_{i+1}) - B(s_i))(B(s_{j+1}) - B(s_j)).$$

When we sum over $i \neq j$, we mean all possible combinations of $i, j \in \{1, \ldots, n\}$ such that $i \neq j$. The summation $i < j$ means analogously all $i, j \in \{1, \ldots, n\}$ such that $i < j$. We consider the expectation of the two sums in the last equality: in the first sum we have that $X(s_i)$ is $\mathcal{F}_{s_i}$-adapted and $B(s_{i+1}) - B(s_i)$ is independent of $\mathcal{F}_{s_i}$ since the increments of a Brownian motion are independent. Hence, $X^2(s_i)$ is independent of $(B(s_{i+1}) - B(s_i))^2$, and

$$\mathbb{E}\left[X^2(s_i)(B(s_{i+1}) - B(s_i))^2\right] = \mathbb{E}\left[X^2(s_i)\right] \mathbb{E}\left[(B(s_{i+1}) - B(s_i))^2\right]$$
$$= \mathbb{E}\left[X^2(s_i)\right](s_{i+1} - s_i).$$

In the last equality we use that $B(s_{i+1}) - B(s_i) \sim \mathcal{N}(0, s_{i+1} - s_i)$.

We prove next that the second sum has zero expectation: since $i < j$, $B(s_{j+1}) - B(s_j)$ will be independent of $B(s_{i+1}) - B(s_i)$ from the independent increment property of Brownian motion. Furthermore, $B(s_{j+1}) - B(s_j)$ is independent of $X(s_i)$ and $X(s_j)$ since they are $\mathcal{F}_{s_i}$- and $\mathcal{F}_{s_j}$-adapted resp., with $\mathcal{F}_{s_i} \subset \mathcal{F}_{s_j}$. Using that the increments of Brownian motion have zero expectation we find

$$\mathbb{E}\left[X(s_i)X(s_j)(B(s_{i+1}) - B(s_i))(B(s_{j+1}) - B(s_j))\right]$$
$$= \mathbb{E}\left[X(s_i)X(s_j)(B(s_{i+1}) - B(s_i))\right]$$
$$\times \mathbb{E}\left[B(s_{j+1}) - B(s_j)\right]$$
$$= 0.$$

We have thus proved the equality.

3.2

The Itô integral is defined as

$$\int_0^t X(s)\,dB(s) = \lim_{n\to\infty} \sum_{i=1}^n X(s_i)\left(B(s_{i+1}) - B(s_i)\right).$$

For each n, the expectation of the sum becomes

$$\mathbb{E}\left[\sum_{i=1}^n X(s_i)\left(B(s_{i+1}) - B(s_i)\right)\right] = \sum_{i=1}^n \mathbb{E}\left[X(s_i)\right]\mathbb{E}\left[B(s_{i+1}) - B(s_i)\right]$$
$$= 0.$$

After taking the limit of the expectation when n goes to infinity we get that the expectation of the Itô integral is zero. The same argumentation as in the solution of Exercise 3.1 gives the variance of the Itô integral.

3.3

We must prove that $aX_s + bY_s$ is adapted and $\mathbb{E}[\int_0^t (aX(s) + bY(s))^2\,ds] < \infty$. But since both $X(s)$ and $Y(s)$ are Itô integrable, they will be in particular adapted and hence dependent on Brownian motion $B(u)$ for all $u \leq s$. This implies that $aX_s + bY_s$ is dependent on Brownian motion $B(u)$ for $u \leq s$ (but not for later times). Hence, $aX(s) + bY(s)$ is $\mathcal{F}_s$-adapted for every s, and therefore an adapted process.

It remains to show that $\mathbb{E}[\int_0^t (aX(s) + bY(s))^2\,ds] < \infty$: it holds[1] that $2abX_sY_s \leq a^2X_s^2 + b^2Y_s^2$, which gives

$$\mathbb{E}[\int_0^t (aX(s)+bY(s))^2\,ds] \leq 2\left(a^2\mathbb{E}[\int_0^t X^2(s)\,ds] + b^2\mathbb{E}[\int_0^t Y^2(s)\,ds]\right) < \infty,$$

where we use the Itô integrability of $X(s)$ and $Y(s)$ to conclude the finiteness of the expectations.

We show that the simple stochastic process $X(s) = 1$ is Itô integrable. Note that all processes of the form $X(s) = f(B(s))$ will be adapted, and therefore in particular $X(s) = 1$ by choosing $f(x) = 1$. Furthermore,

$$\mathbb{E}\left[\int_0^t 1^2\,ds\right] = t < \infty,$$

so the Itô integrability follows. Note that from this we can conclude that all constants are Itô integrable.

---

[1] Prove the basic inequality $2xy \leq x^2 + y^2$ yourself by appealing to the fact that $(x-y)^2 \geq 0$.

3.4

To simplify the notation slightly, define $\Delta B_i := B(s_{i+1}) - B(s_i)$ and $\Delta s_i := s_{i+1} - s_i$. Hence, expectation leaves us with

$$\mathbb{E}\Big[\Big(\sum_{i=1}^{n-1} f''(B(s_i))((\Delta B_i)^2 - \Delta s_i)\Big)^2\Big] = \sum_{i,j=1}^{n-1} \mathbb{E}[f''(B(s_i))f''(B(s_j)) \\ \times ((\Delta B_i)^2 - \Delta s_i)((\Delta B_j)^2 - \Delta s_j)].$$

If $i < j$ then $f''(B(s_i))f''(B(s_j))((\Delta B_i)^2 - \Delta s_i)$ and $(\Delta B_j)^2 - \Delta s_j$ are independent. The terms will therefore have expectation zero in this case since $\mathbb{E}[((\Delta B_j)^2] = \Delta s_j$. Likewise will all terms vanish when $i > j$. Hence, we are left with the terms $i = j$. Using the adaptedness of $f''(B(s_i))$ we find

$$\mathbb{E}\Big[\Big(\sum_{i=1}^{n-1} f''(B(s_i))((\Delta B_i)^2 - \Delta s_i)\Big)^2\Big] = \sum_{i=1}^{n-1} \mathbb{E}[(f''(B(s_i)))^2]\mathbb{E}[(\Delta B_i)^4 \\ - 2(\Delta B_i)^2\Delta s_i + (\Delta s_i)^2]$$

$$= \sum_{i=1}^{n-1} \mathbb{E}[(f''(B(s_i)))^2](3(\Delta s_i)^2 - 2(\Delta s_i)^2 + (\Delta s_i)^2)$$

$$\leq 2 \max_i \Delta s_i \sum_{i=1}^{n-1} \mathbb{E}[(f''(B(s_i)))^2]\Delta s_i.$$

(see Exercise 2.6, where an expression for all moments of a Brownian increment is calculated, and in particular the fourth). Letting $n \to \infty$ it follows $\Delta s_i \downarrow 0$, and therefore we have demonstrated what the exercise asked for.

3.5

We multiply $f''(B(s))$ by 1, and then use the Cauchy–Schwarz inequality with $g(s) = f''(B(s))$ and $h(s) = 1$:

$$\Big(\int_0^t f''(B(s))\,ds\Big)^2 = \Big(\int_0^t f''(B(s))\cdot 1\,ds\Big)^2$$

$$\leq \Big(\Big(\int_0^t f''(B(s))^2\,ds\Big)^{1/2}\Big(\int_0^t 1^2\,ds\Big)^{1/2}\Big)^2$$

$$= t \cdot \int_0^t f''(B(s))^2\,ds.$$

Taking expectation of both sides, we find

$$\mathbb{E}\Big[\big(\int_0^t f''(B(s))\,ds\big)^2\Big] \leq t\mathbb{E}\Big[\int_0^t f''(B(s))^2\,ds\Big] < \infty.$$

3.6

By using the inequality $(a+b)^2 \leq 2a^2 + 2b^2$ twice (see solution to Exercise 3.3 for a hint on how to prove this inequality) we get

$$X^2(t) \leq 2x^2 + 4\left(\int_0^t Y(s)\,dB(s)\right)^2 + 4\left(\int_0^t Z(s)\,ds\right)^2.$$

The Itô isometry yields that

$$\mathbb{E}\left[\left(\int_0^t Y(s)\,dB(s)\right)^2\right] = \mathbb{E}\left[\int_0^t Y^2(s)\,ds\right],$$

which is finite by the Itô integrability of $Y(s)$. Furthermore, from the Cauchy–Schwarz inequality we find

$$\mathbb{E}\left[\left(\int_0^t Z(s)\,ds\right)^2\right] = t \cdot \mathbb{E}\left[\int_0^t Z^2(s)\,ds\right],$$

which also is finite by the Itô integrability of $Z(s)$. Hence, we can conclude that the second moment of X is finite.

3.7

We consider first the Itô integral term. The integrand $Y(s)\partial f(s, X(s))/\partial x$ is an adapted process from the assumptions. From (3.14) we can conclude that $Y(s)\partial f(s, X(s))/\partial x$ is an Itô integrable process. Hence, the Itô integral term exists. From the Itô isometry we see that the second moment is finite.

By arguing as in the solution to Exercise 3.5 we find that the second moment of the ds integral is finite when condition (3.15) holds. This also shows that the integral is finite, of course.

3.8

We first consider (3.14). By identifying all the terms involved, we must show that

$$\mathbb{E}\left[\int_0^t \lambda^2 e^{-2\lambda s} X^2(s)\,ds\right] < \infty.$$

Using the Itô isometry we find

$$\mathbb{E}\left[\int_0^t e^{-2\lambda s} X^2(s)\,ds\right] = \int_0^t e^{-2\lambda s}\mathbb{E}\left[\left(\int_0^s e^{\lambda u}\,dB(u)\right)^2\right]ds$$

$$= \int_0^t e^{-2\lambda s}\int_0^s e^{2\lambda u}\,du\,ds$$

$$= \frac{1}{2\lambda}t - \frac{1}{4\lambda^2}\left(1 - e^{-2\lambda t}\right),$$

which is finite. Hence, condition (3.14) holds.

To prove condition (3.15), we need to show that

$$\mathbb{E}\left[\int_0^t e^{2\lambda s} \cdot e^{-2\lambda s}\,\mathrm{d}s\right] < \infty,$$

which is obviously true.

3.9

The function h is a non-random function, which then must be adapted. From the integrability assumption on h we find that

$$\mathbb{E}\left[\int_0^t h^2(s)\,\mathrm{d}s\right] = \int_0^t h^2(s)\,\mathrm{d}s < \infty,$$

and hence h is Itô integrable.

Let $f(t,x) = h(t)x$. Then $\partial f(t,x)/\partial t = h'(t)x$, $\partial f(t,x)/\partial x = h(t)$ and $\partial^2 f(t,x)/\partial x^2 = 0$. We use Itô's formula for Brownian motion to get

$$\mathrm{d}f(t, B(t)) = h'(t)B(t)\,\mathrm{d}t + h(t)\,\mathrm{d}B(t) + \frac{1}{2} \times 0 \times (\mathrm{d}B(t))^2$$
$$= h'(t)B(t)\,\mathrm{d}t + h(t)\,\mathrm{d}B(t).$$

Integrating both sides yields

$$h(t)B(t) = h(0)B(0) + \int_0^t h'(s)B(s)\,\mathrm{d}s + \int_0^t h(s)\,\mathrm{d}B(s).$$

Rearranging terms proves the desired identity.

3.10

a) By differentiating n times, we find $f^{(n)}(\theta) = \mathbb{E}[e^{\theta X} X^n]$, which immediately implies that $f^{(n)}(0) = \mathbb{E}[X^n]$.

b) We calculate

$$\mathbb{E}\left[\exp(\theta X)\right] = \frac{1}{\sqrt{2\pi a^2}} \int_\mathbb{R} \exp\left(\theta x - \frac{x^2}{2a^2}\right)\,\mathrm{d}x$$
$$= \frac{1}{\sqrt{2\pi a^2}} \int_\mathbb{R} \exp\left(-\frac{1}{2a^2}(x - \theta a^2)^2\right)\,\mathrm{d}x \exp\left(\frac{1}{2}\theta^2 a^2\right)$$
$$= \exp\left(\frac{1}{2}\theta^2 a^2\right).$$

For the opposite result, let $\theta = -2\pi i \xi$ to find

$$\exp\left(-2\pi^2\xi^2 a^2\right) = \int_{\mathbb{R}} \exp\left(-2\pi i \xi x\right)\phi(x)\,\mathrm{d}x = \mathbb{E}\left[\exp\left(-2\pi i X\right)\right],$$

where $\phi(x)$ is the density of X. Hence, the Fourier transform of ϕ is $\exp(-2\pi^2\xi^2 a^2)$, which using inverse Fourier transformation (see, e.g. [25]) implies that $\phi(x) = (2\pi a^2)^{-1/2}\exp(-x^2/2a^2)$, the density function of a normally distributed random variable having mean zero and variance a^2.

c) We calculate the moment generating function of $\int_0^t h(s)\,\mathrm{d}B(s)$. The hint suggests the use of Itô's formula. Let $f(t,x) = \exp(\theta x)$ and put $X(t) = \int_0^t h(s)\,\mathrm{d}B(s)$. Direct differentiation gives $\partial f(t,x)/\partial t = 0$, $\partial f(t,x)/\partial x = \theta\exp(\theta x)$ and $\partial^2 f(t,x)/\partial x^2 = \theta^2\exp(\theta x)$. Hence, by the general Itô formula and $(\mathrm{d}X(t))^2 = h^2(t)\,\mathrm{d}t$ we find

$$\mathrm{d}\exp\left(\theta X(t)\right) = \theta\exp\left(\theta X(t)\right)h(t)\,\mathrm{d}B(t) + \frac{1}{2}\theta^2\exp\left(\theta X(t)\right)h^2(t)\,\mathrm{d}t.$$

Integrating both sides gives

$$\exp\left(\theta X(t)\right) = 1 + \theta\int_0^t \exp\left(\theta X(s)\right)h(s)\,\mathrm{d}B(s)$$
$$+ \frac{1}{2}\theta^2 \int_0^t \exp\left(\theta X(s)\right)h^2(s)\,\mathrm{d}s.$$

Taking the expectation on both sides and using that the Itô integral has zero mean, we get an integral equation for the moment generating function of $\int_0^t h(s)\,\mathrm{d}B(s)$:

$$\mathbb{E}\left[\exp\left(\theta\int_0^t h(s)\,\mathrm{d}B(s)\right)\right] = 1 + \frac{1}{2}\theta^2\int_0^t \mathbb{E}\left[\exp\left(\theta\int_0^s h(u)\,\mathrm{d}B(u)\right)\right]\mathrm{d}s.$$

Defining the function $g(t) = \mathbb{E}\left[\exp(\theta\int_0^t h(s)\,\mathrm{d}B(s))\right]$, and differentiating both sides with respect to t leads to the ordinary differential equations

$$g'(t) = \frac{1}{2}\theta^2 g(t),$$

with initial condition $g(0) = 1$. The reader is urged to prove that the solution to this is

$$\mathbb{E}\left[\exp\left(\theta\int_0^t h(s)\,\mathrm{d}B(s)\right)\right] = g(t) = \exp\left(\frac{1}{2}\theta^2\int_0^t h^2(s)\,\mathrm{d}s\right),$$

which is what we wanted to show.

3.11

We only demonstrate with $X(t) = \mu t + \sigma B(t)$. Let $f(t,x) = S(0)\mathrm{e}^x$, which implies $f(t,X(t)) = S(t)$. Differentiation of f gives $\partial f(t,x)/\partial t = 0$,

$\partial f(t,x)/\partial x = S(0)e^x = f(t,x)$ and $\partial^2 f(t,x)/\partial x^2 = f(t,x)$. Furthermore, since $dX(t) = \mu\,dt + \sigma\,dB(t)$, we have $(dB(t))^2 = \sigma^2\,dt$. Hence, the general Itô formula yields

$$\begin{aligned} dS(t) &= df(t, X(t)) \\ &= f(t, X(t))\,dX(t) + \frac{1}{2} f(t, X(t)) \,(dX(t))^2 \\ &= S(t)\,(\mu\,dt + \sigma\,dB(t)) + \frac{1}{2} S(t) \sigma^2\,dt \\ &= (\mu + 0.5\sigma^2) S(t)\,dt + \sigma S(t)\,dB(t). \end{aligned}$$

We would find exactly the same result when using $X(t) = \sigma B(t)$. In that case you must choose $f(t,x) = S(0)\exp(\mu t + x)$.

By appealing to what we have just calculated, we define $\mu := \alpha - 0.5\sigma^2$, and see that

$$S(t) = S(0)\exp\left(\left(\mu - \frac{1}{2}\sigma^2\right)t + \sigma B(t)\right),$$

solves the stochastic differential equation.

3.12

Define the function $g(t, x_1, x_2) = \exp(x_1 + x_2)$ and let $X_1(t) = B_1(t)$ and $X_2(t) = B_2(t)$. Note that $Y(t) = g(t, X_1(t), X_2(t))$. We differentiate the function g to find $\partial g(t, x_1, x_2)/\partial t = 0$, $\partial g(t, x_1, x_2)/\partial x_i = g(t, x_1, x_2)$ and $\partial g(t, x_1, x_2)/\partial x_i x_j = g(t, x_1, x_2)$ for $i, j = 1, 2$. Moreover, $dX_i(t)dX_j(t) = \delta_{ij} dt$. Hence, by the multi-dimensional Itô formula,

$$\begin{aligned} dY(t) &= dg(t, X_1(t), X_2(t)) \\ &= Y(t)\,dB_1(t) + Y(t)\,dB_2(t) + \frac{1}{2} \times 2 \times Y(t)\,dt \\ &= Y(t)\,dt + Y(t)\,(dB_1(t) + dB_2(t)). \end{aligned}$$

3.13

Define the functions $g_i(t, x_1, \ldots, x_m)$, for $i = 1, \ldots, n$, by

$$g_i(t, x_1, \ldots, x_m) = S_i(0) \exp\left(\left(\alpha_i - \frac{1}{2} \sum_{j=1}^m \sigma_{ij}^2\right) t + \sum_{j=1}^m \sigma_{ij} x_j\right).$$

We have $S_i(t) = g_i(t, B_1(t), \ldots, B_m(t))$. Straightforward differentiation shows

$$\frac{\partial g_i(t, x_1, \ldots, m)}{\partial t} = \left(\alpha_i - \frac{1}{2}\sum_{j=1}^m \sigma_{ij}^2\right) g_i(t, x_1, \ldots, m),$$

$$\frac{\partial g_i(t, x_1, \ldots, m)}{\partial x_k} = \sigma_{ik} g_i(t, x_1, \ldots, m),$$

$$\frac{\partial^2 g_i(t, x_1, \ldots, m)}{\partial x_k \partial x_l} = \sigma_{ik} \sigma_{il} g_i(t, x_1, \ldots, m).$$

Let $X_j(t) = B_j(t)$ in the multi-dimensional Itô formula. This yields

$$\begin{aligned}
\mathrm{d}g_i(t, B_1(t), \ldots, B_m(t)) &= \left(\alpha_i - \frac{1}{2} \sum_{j=1}^m \sigma_{ij}^2 \right) g_i(t, B_1(t), \ldots, B_m(t)) \,\mathrm{d}t \\
&\quad + \sum_{k=1}^m \sigma_{ik} g_i(t, B_1(t), \ldots, B_m(t)) \,\mathrm{d}B_k(t) \\
&\quad + \frac{1}{2} \sum_{k,l=1}^m \sigma_{ik} \sigma_{il} g_i(t, B_1(t), \ldots, B_m(t)) \,\mathrm{d}B_k(t) \,\mathrm{d}B_l(t) \\
&= \alpha_i g_i(t, B_1(t), \ldots, B_m(t)) \,\mathrm{d}t \\
&\quad + \sum_{k=1}^m \sigma_{ik} g_i(t, B_1(t), \ldots, B_m(t)) \,\mathrm{d}B_k(t) \\
&= \alpha_i S_i(t) \,\mathrm{d}t + S_i(t) \sum_{k=1}^m \sigma_{ik} \,\mathrm{d}B_k(t),
\end{aligned}$$

where we have used that $\mathrm{d}B_k(t)\mathrm{d}B_l(t) = \mathrm{d}t$ whenever $k = l$, and zero otherwise. Thus, we have demonstrated what the exercise asked for.

3.14

Letting $n = m = 2$ in Exercise 3.13 we know that

$$S_1(t) = S_1(0) \exp\left(\left(\alpha_1 - \frac{1}{2}(\sigma_{11}^2 + \sigma_{12}^2)\right) t + \sigma_{11} B_1(t) + \sigma_{12} B_2(t) \right),$$

$$S_2(t) = S_2(0) \exp\left(\left(\alpha_2 - \frac{1}{2}(\sigma_{21}^2 + \sigma_{22}^2)\right) t + \sigma_{21} B_1(t) + \sigma_{22} B_2(t) \right).$$

Hence, using the rules of logarithms we find

$$X(t) = \left(\alpha_1 - \frac{1}{2}(\sigma_{11}^2 + \sigma_{12}^2)\right) + \sigma_{11} \Delta B_1(t) + \sigma_{12} \Delta B_2(t),$$

$$Y(t) = \left(\alpha_2 - \frac{1}{2}(\sigma_{21}^2 + \sigma_{22}^2)\right) + \sigma_{21} \Delta B_1(t) + \sigma_{22} \Delta B_2(t),$$

where $\Delta B_i(t) = B_i(t) - B_i(t-1)$ for $i = 1, 2$. Since $B_i(t) - B_i(t-1) \sim \mathcal{N}(0, 1)$ for $i = 1, 2$ and B_1 is independent of B_2, we see that X and Y are normally distributed random variables,

$$X(t) \sim \mathcal{N}\left(\alpha_1 - \frac{1}{2}(\sigma_{11}^2 + \sigma_{12}^2), \sigma_{11}^2 + \sigma_{12}^2\right),$$

$$Y(t) \sim \mathcal{N}\left(\alpha_2 - \frac{1}{2}(\sigma_{21}^2 + \sigma_{22}^2), \sigma_{21}^2 + \sigma_{22}^2\right).$$

From the definition of covariance, we have

$$\begin{aligned}\operatorname{Cov}(X(t), Y(t)) &= \mathbb{E}\left[X(t)Y(t)\right] - \mathbb{E}\left[X(t)\right]\mathbb{E}\left[Y(t)\right] \\ &= \sigma_{11}\sigma_{21}\mathbb{E}\left[(\Delta B_1(t))^2\right] + \sigma_{12}\sigma_{22}\mathbb{E}\left[(\Delta B_2(t))^2\right] \\ &\quad + (\sigma_{11}\sigma_{22} + \sigma_{12}\sigma_{21})\mathbb{E}\left[\Delta B_1(t)\Delta B_2(t)\right] \\ &= \sigma_{11}\sigma_{21} + \sigma_{12}\sigma_{22}.\end{aligned}$$

The correlation follows as

$$\operatorname{corr}(X(t), Y(t)) = \frac{\sigma_{11}\sigma_{21} + \sigma_{12}\sigma_{22}}{\sqrt{\sigma_{11}^2 + \sigma_{12}^2}\sqrt{\sigma_{21}^2 + \sigma_{22}^2}}.$$

It is worth noticing that a usual modelling approach for a two-dimensional geometric Brownian motion is to choose $\sigma_{11} = \sigma_1$, $\sigma_{12} = \sigma_1\rho$, $\sigma_{21} = \sigma_2$ and $\sigma_{22} = 0$. Then

$$\operatorname{corr}(X(t), Y(t)) = \frac{\sigma_1\sigma_2}{\sqrt{\sigma_1^2 + \sigma_1^2\rho^2}\sqrt{\sigma_2^2}} = \frac{\pm 1}{\sqrt{1 + \rho^2}},$$

where we get a positive sign in the numerator when $\sigma_1\sigma_2 > 0$ and negative otherwise. Hence, we can model the correlation through an appropriate choice of ρ.

3.15

a) We describe A in terms of $S(t)$. First, $B(0.5) \in H_1$ is equivalent to $0 < B(0.5, \omega) < \infty$. Multiplying by σ and adding $\mu/2$ yields

$$\mu/2 < \mu/2 + \sigma B(0.5, \sigma) < \infty.$$

Taking the exponential and then multiplying by $S(0)$ gives us

$$S(0)e^{\mu/2} < S(0.5, \omega) < \infty.$$

Similarly,

$$0 < S(2, \omega) < S(0)e^{2\mu}.$$

Introducing the sets

$$\widetilde{H}_1 = \left(S(0)e^{\mu/2}, \infty\right), \quad \widetilde{H}_2 = \left(0, S(0)e^{2\mu}\right),$$

we obtain

$$A = \{\omega \in \Omega \mid S(0.5, \omega) \in \widetilde{H}_1, S(2, \omega) \in \widetilde{H}_2\}.$$

b) If $\omega \in A$, then $B(0.5,\omega) \in (0,\infty)$. Hence, we ask for the probability that $B(0.5) \in (0,\infty)$. Brownian motion is a normal random variable with zero mean, which implies that the probability is 50% since the normal distribution is symmetric.

c) We find a subset H of $\mathbb{R}$ such that $A = \{\omega \in \Omega \mid B(1,\omega) \in H\}$. This will show that $A \in \mathcal{F}_1$. It holds that

$$S(1,\omega) < b \iff S(0)\exp(\mu + \sigma B(1,\omega)) < b$$
$$\iff B(1,\omega) < \frac{1}{\sigma}\left(\ln\frac{b}{S(0)} - \mu\right).$$

Similarly,

$$S(1,\omega) > a \iff B(1,\omega) > \frac{1}{\sigma}\left(\ln\frac{a}{S(0)} - \mu\right).$$

Hence, we find

$$H = \left(\frac{1}{\sigma}\left(\ln\frac{a}{S(0)} - \mu\right), \frac{1}{\sigma}\left(\ln\frac{b}{S(0)} - \mu\right)\right).$$

3.16

Brownian motion starts at zero at time 0, which implies that $B(0,\omega) = 0$ for all $\omega \in \Omega$. For a subset H of $\mathbb{R}$, we have *either* $0 \in H$ *or* $0 \notin H$ (both cannot occur). If $0 \in H$, we have

$$A = \{\omega \in \Omega \mid B(0,\omega) \in H\} = \Omega.$$

In the opposite case, we find

$$A = \{\omega \in \Omega \mid B(0,\omega) \in H\} = \emptyset.$$

Since all subsets $A \in \mathcal{F}_0$ are of the form $A = \{\omega \in \Omega \mid B(0,\omega) \in H\}$, it follows that Ω and $\emptyset$ are the only two possibilities.

3.17

Since k is a constant, it is $\mathcal{F}_s$-adapted. With $X = k$, (3.26) obviously holds for every set $A \in \mathcal{F}_s$. Therefore $k = \mathbb{E}[k \mid \mathcal{F}_s]$.

4.1

Sell N claims and receive $N\widetilde{P}$ that you partly invest in replicating portfolios. Namely, buy N hedging portfolios which costs $H(0) = P(0)$ each. You will have a surplus of $N(\widetilde{P} - P(0))$ from this trade, which you invest in bonds. After one period, you cover the claims by selling the hedges. The hedge is by construction such that it is exactly equal to the size of the claim. Hence, you are left with a profit in bonds $N(\widetilde{P} - P(0))(1+r) > 0$. This is an arbitrage opportunity since we can obtain an arbitrary large profit from a zero net investment.

4.2

Let a be the number of stocks and b the number of bonds in a portfolio. The portfolio will have value $H(T,\omega)$ next time period, and in order to be a hedge for the claim X it must satisfy

$$aS(T,\omega_i) + b(1+r) = X(\omega_i), \ i = 1, 2, 3,$$

where we set the risk-free rate of return on the bond to be r. But we see that this gives *three* equations for only *two* unknowns, which we know does not have any solution in general.

4.3

a) Let a and b be the postion in stock and bond respectively. A superreplicating portfolio must satisfy

$$asu + b \geq x,$$
$$as + b \geq y,$$
$$asd + b \geq 0.$$

Assuming that the last inequality actually holds with equality, we find $b = -asd$, which from the two first inequalities yield

$$a \geq \frac{x}{s(u-d)}, \text{ and } a \geq \frac{y}{s(1-d)}.$$

All superhedging strategies are thus classified as the investment a and b such that $b = -asd$ and

$$a \geq \frac{1}{s}\max\left(\frac{x}{u-d}, \frac{y}{1-d}\right).$$

To proceed, suppose that $x/(u-d) > y/(1-d)$. The price of a superhedge is then

$$H(0) = as + b = as - asd = as(1-d),$$

which implies from the lower bound of a that

$$H(0) \geq x\frac{1-d}{u-d}.$$

Hence, the cheapest superhedging strategy costs $H(0) = x(1-d)/(u-d)$, and is achieved by choosing $a = x/s(u-d)$ and $b = -xd/(u-d)$.

b) A similar calculation for the subhedging strategies reveals (using the assumption $x/(u-d) > y/(1-d)$) that $a \leq y/s(1-d)$ and $b = -asd$. Hence, the most expensive subhedging strategy costs $H(0) = y$ and is achieved by choosing $a = y/s(1-d)$ and $b = -asd$.

c) We have an interval
$$\left(y, \frac{x(1-d)}{u-d}\right),$$
where the initial investments required to superhedge lie to the right, and the initial investments required to subhedge lie to the left of this interval. We now show that if the claim is traded for a price $p \in (y, x(1-d)/(u-d))$, then this price does not allow for any arbitrage opportunities.

Assume by contradiction that the position a, b and c in stock, bond and claim resp., is an arbitrage opportunity. Then $as + b + cp = 0$ and at time T we will have
$$asu + b + cx \geq 0,$$
$$as + b + cy \geq 0,$$
$$asd + b \geq 0,$$
where at least one inequality is strict in order for the expectation to be strictly positive. If just one is strictly positive, and the two others are equal to zero, we have three equations which are linearily independent and thus yielding a solution $a = b = c = 0$, which is contradicting that one inequality is strict. Hence, we must have two or three strict inequalities in order to obtain an arbitrage opportunity.

Let us say that $as + b = 0$, and the other two hold with strict inequality. From the initial investment we find $c = -as(1-d)/p$. Hence, if $a > 0$, then $c < 0$ and we get from $0 < asu + b + cx$ that $-as(u-d) < cx$. Hence,
$$c = \frac{-as(1-d)}{p} = \frac{-as(u-d)(1-d)}{p(u-d)} < \frac{cx(1-d)}{u-d}.$$
Dividing both sides by the negative number c gives us that
$$p > \frac{x(1-d)}{u-d},$$
being a contradiction. On the other hand, if $a < 0$ we know that $c > 0$, and from $0 < as + b + cy$ we find $-as(1-d) < cy$. Thus,
$$c = \frac{-as(1-d)}{p} < c\frac{y}{p}.$$
Dividing by the positive number c gives $p < y$, which again is a contradiction.

The reader is encouraged to work out in a similar way the other cases, which will all contradict the fact that $p \in (y, x(1-d)/(u-d))$. In conclusion, there are no possibilities to construct an arbitrage portfolio consisting of investment in stock, bond and claim when the price belongs to this interval. The reader should note that we have a continuum of prices which all are arbitrage-free, none of which comes from the price of a hedging portfolio.

4.4

Assume, conversely, that $P(t) < S(t) - K\exp(-r(T-t))$ holds. This implies that
$$(S(t) - P(t))e^{r(T-t)} > K.$$
Now, buy N call options costing $NP(t)$, and sell (short) N underlying stocks giving you $NS(t)$. For the surplus $N(S(t) - P(t)) > 0$, you buy bonds. At exercise time, we claim $N\max(S(T) - K, 0)$ from the options. If $S(T) > K$, we receive $N(S(T) - K)$ from the option position, we buy back the stocks for $NS(T)$ and sell our bonds for $N(S(t) - P(t))\exp(r(T-t))$, thus yielding a net position
$$N(S(T) - K) - NS(T) + N(S(t) - P(t))e^{r(T-t)}$$
$$= N((S(t) - P(t))e^{r(T-t)} - K),$$
which is strictly positive. If $S(T) < K$, we get nothing from the options, and our net position after buying back stocks and selling bonds is
$$-NS(T) + N(S(t) - P(t))e^{r(T-t)} = N\left((S(t) - P(t))e^{r(T-t)} - S(T)\right) > 0,$$
since $S(T) < K$. In both cases we have arbitrage, contradicting that the arbitrage-free price of a call can satisfy $P(t) < S(t) - K\exp(-r(T-t))$.

To prove that $P(t) \leq S(t)$, assume, conversely, that $P(t) < S(t)$. If this is the case, sell N options and receive $NP(t)$. For this money, buy N stocks and $N(P(t) - S(t))$ bonds, which is a net investment of zero. At time of exercise, we settle the options, which means that we have to pay $N\max(S(T) - K, 0)$. Selling the stocks and bonds, on the other hand, will give us altogether $NS(T) + N(P(t) - S(t))\exp(r(T-t))$. Hence, we end up with
$$N\min(K, S(T)) + (P(t) - S(t))e^{r(T-t)} > 0,$$
which therefore is an arbitrage.

4.5

Sell N claims and receive $N\widetilde{P}(t)$. Use part of this money to buy N hedges, which will cost you $NH(t)$. Since $\widetilde{P}(t) > H(t)$, you have surplus $N(\widetilde{P}(t) - H(t))$ from the trades that you, for example, can invest in bonds. Now you wait until exercise time T, where the claims are settled by selling the hedges. By definition, one hedge will completely settle one claim, so you will end up with a profit $N(\widetilde{P}(t) - H(t))e^{r(T-t)} > 0$ from a zero investment. You can make the profit arbitrary and large by choosing N arbitrary. Hence, we constructed an arbitrage opportunity.

4.6

We start by calculating the derivatives with respect to z. We find $\partial p(t,z)/\partial z = -p(t,z)z/t$ and

$$\frac{\partial^2 p(t,z)}{\partial z^2} = -\frac{1}{t}p(t,z) - \frac{z}{t}\left(-\frac{z}{t}\right)p(t,z) = \left(\frac{z^2}{t^2} - \frac{1}{t}\right)p(t,z).$$

Differentiation of p with respect to t gives

$$\begin{aligned}\frac{\partial p(t,z)}{\partial t} &= \frac{1}{\sqrt{2\pi}}\left(-\frac{1}{2}\right)t^{-3/2}e^{-z^2/2t} + \frac{1}{\sqrt{2\pi t}}e^{-z^2/2t} \times \left(-\frac{z^2}{2}\right)(-1)t^{-2} \\ &= t^{-1}\left(-\frac{1}{2}\right)p(t,z) + \frac{z^2}{2t}p(t,z) \\ &= \frac{1}{2}\left(\frac{z^2}{t^2} - \frac{1}{t}\right)p(t,z),\end{aligned}$$

which shows that $\partial p(t,z)/\partial t = \frac{1}{2}\partial^2 p(t,z)/\partial z^2$.

4.7

Let us start by showing that

$$Z^{t,x}(s) = x\exp\left(\left(r - \frac{1}{2}\sigma^2\right)(s-t) + \sigma(B(s) - B(t))\right),$$

solves the stochastic differential equation in Thm. 4.4. For this purpose, introduce the function

$$f(s,y) = x\exp\left(\left(r - \frac{1}{2}\sigma^2\right)(s-t) + \sigma y\right),$$

with $y = B^{t,0}(s)$. The notation $B^{t,0}(s)$ means that the Brownian motion starts in state 0 at time t. We want to use Itô's formula to find $df(s, B^{t,0}(s))$. Differentiating f with respect to s and y gives

$$\frac{\partial f(s,y)}{\partial s} = \left(r - \frac{1}{2}\sigma^2\right)f(s,y),$$

$$\frac{\partial f(s,y)}{\partial y} = \sigma f(s,y),$$

$$\frac{\partial^2 f(s,y)}{\partial y^2} = \sigma^2 f(s,y).$$

Note that time now is s, which starts at the fixed time t. Then Itô's formula yields

$$\mathrm{d}f(s,B^{t,0}(s)) = \left(r - \frac{1}{2}\sigma^2\right) f(s,B^{t,0}(s))\,\mathrm{d}s + \sigma f(s,B^{t,0}(s))\,\mathrm{d}B^{t,0}(s)$$
$$+ \frac{1}{2}\sigma^2 f(s,B^{t,0}(s))\,\mathrm{d}s$$
$$= rf(s,B^{t,0}(s))\,\mathrm{d}s + \sigma f(s,B^{t,0}_s)\,\mathrm{d}B^{t,0}(s).$$

If we now write up the integral version, where we start the integration from t and let it go up to s, we find (using that $B^{t,0}(t) = 0$)

$$f(s,B^{t,0}(s)) - f(t,0) = \int_t^s rf(u,B^{t,0}(u))\,\mathrm{d}u + \int_t^s \sigma f(u,B^{t,0}(u))\,\mathrm{d}B(u).$$

Letting $B^{t,0}(s) := B(s) - B(t)$, we have a Brownian motion running from time t with initial state 0. Then $Z^{t,x}(u) = f(u,B^{t,0}(u))$ and $f(t,0) = x$ for all $u \in [t,s]$, and we obtain

$$Z^{t,x}(s) = x + \int_t^s rZ^{t,x}(u)\,\mathrm{d}u + \int_t^s \sigma Z^{t,x}(u)\,\mathrm{d}B(u),$$

which is the dynamics referred to in Thm. 4.4. Note that we do not need to put superscripts that indicate the initial state of the Brownian motion in the Itô integral since only the *increments* matter and not the starting point of the process itself.

Considering $\ln Z^{t,x}(T)$, we find

$$\ln Z^{t,x}(T) = \ln x + \left(r - \frac{1}{2}\sigma^2\right)(T-t) + \sigma(B(T) - B(t))$$

which from the definition of Brownian motion is a normally distributed random variable with mean $\ln x + (r - \sigma^2/2)(T-t)$ and variance $\sigma^2(T-t)$. Hence, the density function of $\ln Z^{t,x}(T)$ is

$$p(t,z) = \frac{1}{\sqrt{2\pi\sigma^2(T-t)}} \exp\left(-\frac{(z - \ln x - (r - \sigma^2/2)(T-t))^2}{2\sigma^2(T-t)}\right).$$

We can therefore express $C(t,x)$ as

$$C(t,x) = \mathrm{e}^{-r(T-t)} \int_{-\infty}^{\infty} f(\mathrm{e}^z) p(t,z;x)\,\mathrm{d}z.$$

We proceed with finding the derivatives of $C(t,x)$. Since the density p is the only function that depends on x in the expression for $C(t,x)$, we find these derivatives from differentiating p. Hence,

$$\frac{\partial p(t,z)}{\partial t} = p(t,z)\Big\{\frac{1}{2(T-t)} - \frac{(r-\sigma^2/2)(z - \ln x - (r-\sigma^2/2)(T-t))}{\sigma^2(T-t)}$$
$$- \frac{(z - \ln x - (r-\sigma^2/2)(T-t))^2}{2\sigma^2(T-t)^2}\Big\},$$

$$\frac{\partial p(t,z)}{\partial x} = p(t,z)\left\{\frac{z - \ln x - (r - \sigma^2/2)(T-t)}{x\sigma^2(T-t)}\right\},$$

$$\frac{\partial^2 p(t,z)}{\partial x^2} = p(t,z)\left\{\frac{(z - \ln x - (r - \sigma^2/2)(T-t))^2}{x^2\sigma^4(T-t)^2}\right.$$
$$\left. - \frac{z - \ln x - (r - \sigma^2/2)(T-t)}{x^2\sigma^2(T-t)}\right\}.$$

Whence, we see that

$$\frac{\partial p(t,z)}{\partial t} + rx\frac{\partial p(t,z)}{\partial x} + \frac{1}{2}\sigma^2 x^2 \frac{\partial^2 p(t,z)}{\partial x^2} = 0.$$

After commuting integration and differentiation, we find

$$\frac{\partial C(t,x)}{\partial t} = rC(t,x) + e^{-r(T-t)}\frac{\partial}{\partial t}\int_{\mathbb{R}} f(e^z) p(t,z)\,dz$$

$$= rC(t,x) + e^{-r(T-t)}\int_{\mathbb{R}} f(e^z)\frac{\partial p(t,z)}{\partial t}\,dz$$

$$= rC(t,x) + e^{-r(T-t)}\int_{\mathbb{R}} f(e^z)\left\{-rx\frac{\partial p(t,z)}{\partial x} - \frac{1}{2}\sigma^2 x^2 \frac{\partial^2 p(t,z)}{\partial x^2}\right\}dz$$

$$= rC(t,x) - rx\frac{\partial C(t,x)}{\partial x} - \frac{1}{2}\sigma^2 x^2 \frac{\partial^2 C(t,x)}{\partial x^2}.$$

We solve the exercise by observing that

$$C(T,x) = \mathbb{E}_Q\left[f(S^{T,x}(T)\right] = f(x).$$

4.8

First recall the Black & Scholes formula in Thm. 4.6 for a call option:

$$P(t;T,K,r,S(t),\sigma) = S(t)\Phi(d_1) - Ke^{-r(T-t)}\Phi(d_2),$$

with $d_1 = d_2 + \sigma\sqrt{T-t}$ and

$$d_2 = \frac{\ln(S(t)/K) + (r - \sigma^2/2)(T-t)}{\sigma\sqrt{T-t}}.$$

Let $\sigma \downarrow 0$. Then

$$\lim_{\sigma \downarrow 0} d_1 = \lim_{\sigma \downarrow 0} \frac{\ln(S(t)/K) + r(T-t)}{\sigma\sqrt{T-t}} + \frac{1}{2}\sigma\sqrt{T-t}$$
$$= \begin{cases} \infty, & \ln(S(t)/K) + r(T-t) < 0, \\ -\infty, & \ln(S(t)/K) + r(T-t) > 0, \end{cases}$$

$$\lim_{\sigma \downarrow 0} d_2 = \lim_{\sigma \downarrow 0} \frac{\ln(S(t)/K) + r(T-t)}{\sigma\sqrt{T-t}} - \frac{1}{2}\sigma\sqrt{T-t}$$
$$= \begin{cases} \infty, & \ln(S(t)/K) + r(T-t) > 0, \\ -\infty, & \ln(S(t)/K) + r(T-t) < 0. \end{cases}$$

Hence, with $\ln(S(t)/K) + r(T-t) > 0$ we find
$$\lim_{\sigma \downarrow 0} P(t; T, K, r, S(t), \sigma) = S(t)\Phi(\infty) - Ke^{-r(T-t)}\Phi(\infty) = S(t) - Ke^{-r(t-t)}.$$

On the other hand, when $\ln(S(t)/K) + r(T-t) < 0$,
$$\lim_{\sigma \downarrow 0} P(t; T, K, r, S(t), \sigma) = S(t)\Phi(-\infty) - Ke^{-r(T-t)}\Phi(-\infty) = 0.$$

We also must include the special case when $\ln(S(t)/K) + r(T-t) = 0$, for which we will get
$$\lim_{\sigma \downarrow 0} P(t; T, K, r, S(t), \sigma) = S(t)\Phi(0) - Ke^{-r(T-t)}\Phi(0) = \frac{1}{2}\left(S(t) - Ke^{-r(t-t)}\right)$$

To prove that $P(t; T, K, r, S(t), \sigma)$ is increasing in T it is sufficient to show that $\partial P(t; T, K, r, S(t), \sigma)/\partial T > 0$. It is easily seen that
$$\frac{\partial d_1}{\partial T} = -\frac{\ln(S(t)/K)}{2\sigma(T-t)^{3/2}} + \frac{r+\sigma^2/2}{2\sigma\sqrt{T-t}},$$
$$\frac{\partial d_2}{\partial T} = -\frac{\ln(S(t)/K)}{2\sigma(T-t)^{3/2}} + \frac{r-\sigma^2/2}{2\sigma\sqrt{T-t}}.$$

Hence,
$$\frac{\partial P(t)}{\partial T} = S(t)\Phi'(d_1)\frac{\partial d_1}{\partial T} - K(-r)e^{-r(T-t)}\Phi(d_2) + Ke^{-r(T-t)}\Phi'(d_2)\frac{\partial d_2}{\partial T}$$
$$= rKe^{-r(T-t)}\left(r\Phi(d_2) + \frac{\sigma\phi(d_2)}{2\sqrt{T-t}}\right)$$
$$> 0.$$

This shows that the option price is an increasing function of exercise time. The proof that $P(t)$ is increasing in volatility is similar, and left to the reader.
Since $\partial d_1/\partial K = -1/\sigma K\sqrt{T-t} = \partial d_2/\partial K$, we have
$$\frac{\partial P(t)}{\partial K} = -\frac{S(t)\phi(d_1)}{\sigma K\sqrt{T-t}} - e^{-r(T-t)}\Phi(d_2) + \frac{Ke^{-r(T-t)}\phi(d_2)}{\sigma K\sqrt{T-t}}$$
$$= -e^{-r(T-t)}\Phi(d_2)$$
$$< 0,$$

which shows that $P(t)$ is decreasing as a function of K.

4.9

The payoff function from the digital option is

$$f(S(T)) = \begin{cases} 1, & S(T) > S(0), \\ 0, & \text{otherwise}, \end{cases}$$

that is, $f(x) = 1_{\{x > S(0)\}}$. The price is therefore given by $P(t) = C(t, S(t))$, where

$$C(t, x) = e^{-r(T-t)} \mathbb{E}_Q \left[1_{\{S^{t,x}(T) > S(0)\}} \right].$$

We calculate this expectation: from theory we know that

$$S^{t,x}(T) = x \exp\left(\left(r - \frac{1}{2}\sigma^2\right)(T-t) + \sigma(W(T) - W(t))\right),$$

where $W(t)$ is a Brownian motion with respect to Q. Hence, letting $Y \sim \mathcal{N}(0,1)$, we have

$$\mathbb{E}_Q \left[1_{\{S^{t,x}(T) > S(0)\}} \right] = \mathbb{E}_Q \left[1_{\{x \exp((r-\sigma^2/2)(T-t) + \sigma\sqrt{T-t}Y) > S(0)\}} \right]$$
$$= \frac{1}{\sqrt{2\pi}} \int_{-\infty}^{d_2} e^{-z^2/2} \, dz$$
$$= \Phi(d_2),$$

with d_2 as in the Black & Scholes formula, Thm. 4.6. Hence, the price of a call is given by the function $C(t,x) = \exp(-r(T-t))\Phi(d_2)$.

4.10

The price of a call and put is, respectively,

$$P^c(t) = e^{-r(T-t)} \mathbb{E}_Q \left[\max(0, S(T) - K) \right],$$
$$P^p(t) = e^{-r(T-t)} \mathbb{E}_Q \left[\max(0, K - S(T)) \right].$$

Using the hint given in the exercise, we find

$$P^c(t) = e^{-r(T-t)} \mathbb{E}_Q \left[\max(0, S(T) - K) \mid \mathcal{F}_t \right]$$
$$= e^{-r(T-t)} \mathbb{E}_Q \left[(S(T) - K) \mid \mathcal{F}_t \right] + e^{-r(T-t)} \mathbb{E}_Q \left[\max(0, K - S(T)) \mid \mathcal{F}_t \right]$$
$$= e^{rt} \mathbb{E}_Q \left[e^{-rT} S(T) \mid \mathcal{F}_t \right] - e^{-r(T-t)} K + P^p(t)$$
$$= e^{rt} \times e^{-rt} S(t) - Ke^{-r(T-t)} + P^p(t)$$
$$= S(t) - Ke^{-r(T-t)} + P^p(t).$$

In the second equality we used that the conditional expectation of a constant is the constant itself (Exercise 3.17), while the third follows from the martingale property of $e^{-rt}S(t)$ with respect to Q.

4.11

Recall from Thm. 4.7 the delta of a call option given as $\partial P^c(t)/\partial S(t) = \Phi(d_1)$ with d_1 given in Thm. 4.6 and $P^c(t)$ denoting the price of a call option at time t. Using the notation $P^p(t)$ for the price of the corresponding put option at time t, the put–call parity derived in Exercise 4.10 gives

$$\frac{\partial P^c(t)}{\partial S(t)} - \frac{\partial P^p(t)}{\partial S(t)} = 1.$$

Thus, the hedge ratio for a put option is $\partial P^p(t)/\partial S(t) = \Phi(d_1) - 1$.

Note that the ratio is negative, which implies that we should be short in the stock when hedging a put option.

4.12

From Subsect. 4.3.4 we have that the delta of an option paying $f(S(T))$ is given by

$$\frac{\partial C(t,x)}{\partial x} = e^{-r(T-t)} \mathbb{E}\left[g(t,x,Z^{t,x}(T))\right],$$

when using the density approach (letting $x = S(t)$). The function g will in the case of a digital be

$$g(t,x,z) = 1_{\{z>K\}} \frac{\ln z - \ln x - (r-\sigma^2/2)(T-t)}{x\sigma^2(T-t)}.$$

Since

$$Z^{t,x}(T) = x\exp\left(\left(r - \frac{1}{2}\sigma^2\right)(T-t) + \sigma(B(T)-B(t))\right),$$

we calculate

$$g(t,x,Z^{t,x}(T)) = \sigma\sqrt{T-t}\,Y,$$

for a random variable $Y \sim \mathcal{N}(0,1)$. Hence,

$$\frac{\partial C(t,x)}{\partial x} = e^{-r(T-t)}\sigma\sqrt{T-t}\,\mathbb{E}\left[Y 1_{\{x\exp((r-\sigma^2/2)(T-t)+\sigma\sqrt{T-t}Y)>K\}}\right]$$

$$= e^{-r(T-t)} \frac{1}{x\sqrt{2\pi\sigma^2(T-t)}} \int_u^\infty y e^{y^2/2}\,dy$$

$$= e^{-r(T-t)-u^2/2} \frac{1}{x\sqrt{2\pi\sigma^2(T-t)}},$$

where

$$u = \frac{\ln(x/K) - (r-\sigma^2/2)(T-t)}{\sigma\sqrt{T-t}}.$$

4.13

The probability Q is defined by $Q(A) = \mathbb{E}\left[1_A M(T)\right]$, where
$$M(T) = \exp\left(-\frac{\alpha - r}{\sigma} B(T) - \frac{1}{2} \frac{(\alpha - r)^2}{\sigma^2} T\right).$$

The random variable $M(T)$ is positive, which implies that $Q(A) \geq 0$ for all events $A \subset \Omega$. Furthermore, since $B(T)$ is a normally distributed random variable with zero mean and variance T,

$$\begin{aligned} Q(\Omega) &= \mathbb{E}\left[1_\Omega M(T)\right] \\ &= \mathbb{E}\left[M(T)\right] \\ &= \exp\left(-\frac{1}{2} \frac{(\alpha - r)^2}{\sigma^2}\right) \mathbb{E}\left[\exp\left(-\frac{\alpha - r}{\sigma} B(T)\right)\right] \\ &= 1. \end{aligned}$$

Hence, it follows that $0 \leq Q(A) \leq 1$ and $Q(\Omega) = 1$. Assume next that A and B are two disjoint events. It is left to the reader to argue that $1_{A \cup B} = 1_A + 1_B$. Therefore,

$$\begin{aligned} Q(A \cup B) &= \mathbb{E}\left[1_{A \cup B} M(T)\right] \\ &= \mathbb{E}\left[1_A M(T)\right] + \mathbb{E}\left[1_B M(T)\right] \\ &= Q(A) + Q(B). \end{aligned}$$

Finally, if A_i, $i = 1, 2, \ldots$ are events such that $A_i \cap A_j = \emptyset$ for every $i \neq j$, we have[2]

$$\begin{aligned} Q(\cup_{i=1}^\infty A_i) &= \mathbb{E}\left[1_{\cup_{i=1}^\infty A_i} M(T)\right] \\ &= \mathbb{E}\left[\sum_{i=1}^\infty 1_{A_i} M(T)\right] \\ &= \sum_{i=1}^\infty \mathbb{E}\left[1_{A_i} M(T)\right] \\ &= \sum_{i=1}^\infty Q(A_i). \end{aligned}$$

Thus, we have proved that Q is a probability on Ω.

---

[2] In these calculations we commute expectation and summation. Strictly speaking, we must verify that we can move the infinite sum outside the expectation, but we leave this for the mathematically inclined reader.

4.14

The dynamics of $S(t)$ with respect to $\mathcal{Q}$ is

$$dS(t) = rS(t)\,dt + \sigma S(t)\,dW(t).$$

Hence, Itô's formula gives

$$\begin{aligned}
d(e^{-rt}S(t)) &= -r(e^{-rt}S(t))\,dt + e^{-rt}\,dS(t) \\
&= -r(e^{-rt}S(t))\,dt + r(e^{-rt}S(t))\,dt + \sigma(e^{-rt}S(t))\,dW(t) \\
&= \sigma(e^{-rt}S(t))\,dW(t)\,.
\end{aligned}$$

From the martingale representation theorem we conclude that $e^{-rt}S(t)$ is a $\mathcal{Q}$-martingale.

4.15

We use the Itô formula for Brownian motion with the function $f(t,x) = \exp(-\lambda x - \lambda^2 t/2)$. We find $\partial f(t,x)/\partial t = -\lambda^2 f(t,x)/2$, $\partial f(t,x)/\partial x = -\lambda f(t,x)$ and $\partial^2 f(t,x)/\partial x^2 = \lambda^2 f(t,x)$. Hence,

$$\begin{aligned}
dM(t) &= -\frac{1}{2}\lambda^2 M(t)\,dt - \lambda M(t)\,dB(t) + \frac{1}{2}\lambda^2 M(t)\,dt \\
&= -\lambda M(t)\,dB(t).
\end{aligned}$$

It follows from the martingale representation theorem that $M(t)$ is a martingale.

4.16

We calculate for a $\theta \in \mathbb{R}$,

$$\begin{aligned}
\mathbb{E}_{\mathcal{Q}}\left[e^{\theta W(t)}\right] &= \mathbb{E}_{\mathcal{Q}}\left[M(T)e^{\theta W(t)}\right] \\
&= \mathbb{E}\left[\exp\left(-\lambda B(T) + \theta B(t) + \theta \lambda t - 0.5\lambda^2 T\right)\right] \\
&= \exp\left(\theta \lambda t - \frac{1}{2}\lambda^2 T\right)\mathbb{E}\left[\exp\left(\theta B(t) - \lambda(B(T) - B(t)) - \lambda B(t)\right)\right] \\
&= \exp\left(\theta \lambda t - \frac{1}{2}\lambda^2 T\right)\mathbb{E}\left[\exp\left(-\lambda(B(T) - B(t))\right)\right] \\
&\quad \times \mathbb{E}\left[\exp\left((\theta - \lambda)B(t)\right)\right] \\
&= \exp\left(\frac{1}{2}\theta^2 t\right).
\end{aligned}$$

From this we can conclude that $W(t) \sim \mathcal{N}(0,t)$ with respect to the probability $\mathcal{Q}$.

4.17

We have $H(t) = a(t)S(t) + b(t)R(t)$ and from the self-financing hypothesis
$$dH(t) = a(t)\,dS(t) + b(t)\,dR(t).$$
Thus, $H(t)$ is a semimartingale with respect to $\mathcal{Q}$ since
$$dS(t) = rS(t)\,dt + \sigma S(t)\,dW(t).$$
Itô's formula with $X(t) = H(t)$ and $f(t,x) = xe^{-rt}$ yields that
$$\begin{aligned}d(e^{-rt}H(t)) &= -re^{-rt}H(t)\,dt + e^{-rt}\,dH(t)\\ &= -re^{-rt}\left(a(t)S(t)+b(t)R(t)\right)dt + b(t)e^{-rt}rR(t)\,dt\\ &\quad + a(t)\left(re^{-rt}S(t)\,dt + \sigma e^{-rt}S(t)\,dW(t)\right)\\ &= \sigma a(t)e^{-rt}S(t)\,dW(t).\end{aligned}$$

We conclude from the martingale representation theorem that the discounted value of the portfolio is a martingale.

4.18

a) The price of the chooser option is
$$\begin{aligned}P(0) &= e^{-rT}\mathbb{E}_\mathcal{Q}\left[\max(0, S(T)-K)\right] + e^{-rt}\mathbb{E}_\mathcal{Q}\left[\max(0, Ke^{-r(T-t)} - S(t))\right]\\ &= P^c(0;K,T) + P^p(0;Ke^{-r(T-t)},t).\end{aligned}$$
From the put–call parity (see Exercise 4.10)
$$\begin{aligned}P^p(0;Ke^{-r(T-t)},t) &= P^c(0;Ke^{-r(T-t)},t) - S(0) + Ke^{-r(T-t)} \times e^{-rt}\\ &= P^c(0;Ke^{-r(T-t)},t) - S(0) + Ke^{-rT}.\end{aligned}$$
Hence, the price of the chooser option becomes
$$P(0) = P^c(0;K,T) + P^c(0;Ke^{-r(T-t)},t) - S(0) + Ke^{-rT}.$$
The price of the two call options can be expressed by the Black & Scholes formula. It is straightforward to use the Black & Scholes formula to find the price $P^c(0;K,T)$, while $P^c(0;Ke^{-r(T-t)},t)$ becomes
$$\begin{aligned}P^c(0;Ke^{-r(T-t)},t) &= S(0)\Phi(d_1) - Ke^{-r(T-t)} \times e^{-rt}\Phi(d_2)\\ &= S(0)\Phi(d_1) - Ke^{-rT}\Phi(d_2),\end{aligned}$$
where
$$d_1 = \frac{\ln(S(0)/(Ke^{-r(T-t)})) + (r+\sigma^2/2)t}{\sigma\sqrt{t}},$$
$$d_2 = \frac{\ln(S(0)/(Ke^{-r(T-t)})) + (r-\sigma^2/2)t}{\sigma\sqrt{t}}.$$

b) The payoff function from the chooser option is

$$X = \max\left(S(T) - K, 0\right) 1_{\{P^c(t) \geq P^p(t)\}} + \max\left(K - S(T), 0\right) 1_{\{P^c(t) < P^p(t)\}}.$$

Adding and subtracting

$$\max(K - S(T), 0) 1_{\{P^c(t) \geq P^p(t)\}},$$

to X yields

$$X = \max(K - S(T), 0) + \max\left(S(T) - K, 0\right) 1_{\{P^c(t) \geq P^p(t)\}}.$$

Hence, the price of the chooser option becomes

$$\begin{aligned} P(0) &= e^{-rT} \mathbb{E}_\mathcal{Q}[X] \\ &= e^{-rT} \mathbb{E}_\mathcal{Q}\left[\max(K - S(T), 0)\right] \\ &\quad + e^{-rT} \mathbb{E}_\mathcal{Q}\left[\max\left(S(T) - K, 0\right) 1_{\{P^c(t) \geq P^p(t)\}}\right] \\ &= P^p(0) + e^{-rT} \mathbb{E}_\mathcal{Q}\left[\max\left(S(T) - K, 0\right) 1_{\{P^c(t) \geq P^p(t)\}}\right]. \end{aligned}$$

We analyse the expectation in the same way as in Subsect. 4.5.1 by using the law of double expectation, adaptedness of option prices and the martingale property of $e^{-rt}S(t)$ with respect to $\mathcal{Q}$. This will lead to

$$P(0) = P^p(0) + e^{-rt}\mathbb{E}_\mathcal{Q}\left[\max\left(0, S(t) - Ke^{-r(T-t)}\right)\right].$$

We remark that the last term is the price of a call option at time 0 with strike $K\exp(-r(T-t))$ and exercise time t.

4.19

a) The owner of the forward contract will pay $F(t,T)$ at time of exercise T, and receive a stock with value $S(T)$. If the owner sells the stock immediately after receiving it, the net income (or loss) will be $S(T) - F(t,T)$. Hence, the payoff from the forward contract is $X = S(T) - F(t,T)$. Since X is $\mathcal{F}_T$-adapted (it is not reasonable that $F(t,T)$ depends on S after delivery T), it is a contingent claim, and according to Thm. 4.13, the price $P(t)$ at time t is

$$P(t) = e^{-r(T-t)}\mathbb{E}_\mathcal{Q}\left[S(T) - F(t,T) \mid \mathcal{F}_t\right].$$

b) Since $F(t,T)$ is $\mathcal{F}_t$-adapted, (3.28) tells us that

$$\mathbb{E}\left[F(t,T) \mid \mathcal{F}_t\right] = F(t,T).$$

Hence, since $F(t,T)$ is such that $P(t) = 0$, we find

$$0 = e^{-r(T-t)}\mathbb{E}_\mathcal{Q}\left[S(T) - F(t,T) \mid \mathcal{F}_t\right] \iff F(t,T) = \mathbb{E}_\mathcal{Q}\left[S(T) \mid \mathcal{F}_t\right].$$

We observe that $F(t,T)$ is a martingale with respect to $\mathcal{Q}$. Using that $e^{-rt}S(t)$ is a martingale with respect to $\mathcal{Q}$, we find

$$\begin{aligned} F(t,T) &= \mathbb{E}_\mathcal{Q}\left[S(T)\,|\,\mathcal{F}_t\right] \\ &= e^{rT}\mathbb{E}_\mathcal{Q}\left[e^{-rT}S(T)\,|\,\mathcal{F}_t\right] \\ &= e^{rT}e^{-rt}S(t) \\ &= e^{r(T-t)}S(t). \end{aligned}$$

We conclude that the present value of the forward price $F(t,T)$ is equal to the stock price.

c) We use Itô's formula with $f(t,x) = e^{r(T-t)}x$ and $X(t) = S(t)$. Since $\partial f(t,x)/\partial t = -rf(t,x)$, $\partial f(t,x)/\partial x = e^{r(T-t)}$, $\partial^2 f(t,x)/\partial x^2 = 0$ and

$$dS(t) = rS(t)\,dt + \sigma S(t)\,dW(t),$$

we get

$$dF(t,T) = -rF(t,T)\,dt + e^{r(T-t)}\,dS(t) = \sigma F(t,T)\,dW(t).$$

Observe also that $F(T,T) = S(T)$, which is natural since immediate delivery of the spot must have the same value as the spot itself.

d) The call option on the forward with strike K and exercise $\tau < T$ has payoff

$$X = \max\left(F(\tau,T) - K, 0\right).$$

We rewrite X,

$$X = \max\left(e^{r(T-\tau)}S(\tau) - K, 0\right) = e^{r(T-\tau)}\max\left(S(\tau) - K, 0\right).$$

Therefore, we find the price of the option at time t by Thm. 4.13 to be

$$\begin{aligned} P(t) &= e^{-r(\tau-t)}\mathbb{E}_\mathcal{Q}\left[e^{r(T-\tau)}\max\left(S(\tau) - Ke^{-r(T-\tau)}, 0\right)\right] \\ &= e^{r(T-\tau)}\left\{e^{-r(\tau-t)}\mathbb{E}_\mathcal{Q}\left[\max\left(S(\tau) - K, 0\right)\right]\right\}. \end{aligned}$$

The expression in the curly brackets is the price of a call option on the stock S with strike price $Ke^{-r(T-\tau)}$ and exercise at time τ. This price can be found using the Black & Scholes formula in Thm. 4.6. We get

$$\begin{aligned} P(t) &= e^{r(T-\tau)}\left\{S(t)\Phi(d_1) - (Ke^{-r(T-\tau)})e^{-r(\tau-t)}\Phi(d_2)\right\} \\ &= e^{-r(\tau-t)}\left\{F(t,T)\Phi(d_1) - K\Phi(d_2)\right\}, \end{aligned}$$

where $d_1 = d_2 + \sigma\sqrt{\tau - t}$ and

$$d_2 = \frac{\ln(S(t)/Ke^{-r(T-\tau)}) + (r - \sigma^2/2)(\tau - t)}{\sigma\sqrt{\tau - t}}.$$

A simple rewriting of d_2 gives,

$$d_2 = \frac{\ln(F(t,T)/K) - \sigma^2(\tau - t)/2}{\sigma\sqrt{\tau - t}}.$$

This pricing formula has the name "Black-76", and was first presented in [10].

4.20

We substitute $\mathbf{B}(t)$ with the $\mathcal{Q}$-Brownian motion (at least when using the density approach for deriving the delta) $\mathbf{W}(t) = \mathbf{B}(t) + \lambda t$ in (4.28):

$$dS_i(t) = \alpha_i S_i(t)\, dt + S_i(t) \sum_{j=1}^n \sigma_{ij}\, dB_j(t)$$

$$= \left(\alpha_i - \sum_{j=1}^n \sigma_{ij}\lambda_j\right) S_i(t)\, dt + S_i(t) \sum_{j=1}^n \sigma_{ij}\, dW_j(t).$$

But $\alpha_i - \sum_{j=1}^n \sigma_{ij}\lambda_j$ is the ith component of the vector $\alpha - \Sigma\lambda$, which we find to be

$$\alpha - \Sigma\lambda = \alpha - \Sigma\Sigma^{-1}(\alpha - r\mathbf{1}) = r\mathbf{1}.$$

Then the dynamics (4.32) follows.

4.21

a) From Sect. 4.7, we must choose $\mathcal{Q}$ as the probability

$$\mathcal{Q}(A) = \mathbb{E}\left[1_A M(T)\right],$$

for every $A \subset \Omega$ and with

$$M(t) = \exp\left(-\lambda' \mathbf{B}(t) - \frac{1}{2}\lambda'\lambda t\right).$$

The vector λ must be chosen as

$$\lambda = \Sigma^{-1}(\alpha - r\mathbf{1}).$$

The stock price dynamics implies that

$$\alpha = \begin{bmatrix} \alpha_1 \\ \alpha_2 \end{bmatrix},\ \Sigma = \begin{bmatrix} \sigma_1 & \sigma\rho \\ 0 & \sigma_2 \end{bmatrix}.$$

Hence,

150   A Solutions to Selected Exercises

$$\lambda = \begin{bmatrix} \frac{\alpha_1-r}{\sigma_1} - \rho\frac{\alpha_2-r}{\sigma_2} \\ \frac{\alpha_2-r}{\sigma_2} \end{bmatrix},$$

and

$$M(t) = \exp\left(-\left(\frac{\alpha_1-r}{\sigma_1} - \rho\frac{\alpha_2-r}{\sigma_2}\right)B_1(t) - \frac{\alpha_2-r}{\sigma_2}B_2(t)\right.$$
$$\left. - \frac{1}{2}\left(\frac{\alpha_1-r}{\sigma_1} - \rho\frac{\alpha_2-r}{\sigma_2}\right)^2 - \frac{1}{2}\left(\frac{\alpha_2-r}{\sigma_2}\right)^2\right).$$

This characterizes the equivalent martingale measure $\mathcal{Q}$ for the current two-dimensional geometric Brownian motion.
The Brownian motion with respect to $\mathcal{Q}$ we find as

$$W_1(t) = B_1(t) + \left(\frac{\alpha_1-r}{\sigma_1} - \rho\frac{\alpha_2-r}{\sigma_2}\right)t,$$
$$W_1(t) = B_2(t) + \frac{\alpha_2-r}{\sigma_2}.$$

From Exercise 4.20, we then have

$$dS_1(t) = rS_1(t)\,dt + \sigma_1 S_1(t)\left(dW_1(t) + \rho dW_2(t)\right),$$
$$dS_2(t) = rS_2(t)\,dt \sigma_2 S_2(t)\,dW_2(t).$$

b) The explicit forms of $S_1(T)$ and $S_2(T)$ are (use the multi-dimensional Itô formula to show this)

$$S_1(T) = S_1(0)\exp\left(\left(r - \frac{1}{2}\sigma_1^2(1+\rho^2)\right)T + \sigma_1(W_1(T) + \rho W_2(T))\right),$$
$$S_2(T) = S_2(0)\exp\left(\left(r - \frac{1}{2}\sigma_2^2\right)T + \sigma_2 W_2(T)\right).$$

If we suppose that $S_1(T) - S_2(T)$ is close to normally distributed, we can find the mean and variance of that distribution by calculating the mean and variance of this difference. Denote by m_T and v_T^2 the mean and variance of $S_1(T) - S_2(T)$ resp. Straightforward calculations then reveal that

$$m_T := \mathbb{E}_\mathcal{Q}[S_1(T) - S_2(T)] = e^{rT}(S_1(0) - S_2(0)),$$
$$v_T^2 := \mathbb{E}_\mathcal{Q}\left[(S_1(T) - S_2(T) - m_T)^2\right]$$
$$= e^{2rT}\left\{S_1^2(0)\exp(\sigma_1^2(1+\rho^2)T) + S_2^2(0)\exp(\sigma_2^2 T) - (S_1(0) - S_2(0))^2\right.$$
$$\left. - 2S_1(0)S_2(0)\exp\left(\frac{1}{2}\sigma_1^2(1-\rho^2)T + \sigma_1\sigma_2\rho T + \frac{1}{2}\sigma_2^2(1-\rho^2)T\right)\right\}.$$

Hence, we have that

$$S_1(T) - S_2(T) \approx m_T + v_T Y,$$

for a random variable $Y \sim \mathcal{N}(0,1)$, and where the approximate equality is in the sense of distributions. Therefore, the price $P(0)$ can be approximated as

$$\begin{aligned}
P(0) &= e^{-rT} \mathbb{E}_{\mathcal{Q}} \left[\max(S_1(T) - S_2(T) - K, 0) \right] \\
&\approx e^{-rT} \mathbb{E}_{\mathcal{Q}} \left[\max(m_T + v_T Y - K, 0) \right] \\
&= e^{-rT} \frac{1}{\sqrt{2\pi}} \int_{m_T + v_T y > K} (m_T + v_T y - K) e^{-y^2/2} \, dy \\
&= e^{-rT} \frac{1}{\sqrt{2\pi}} \left\{ \int_{-\infty}^{d} (m_T - K) e^{-y^2/2} \, dy + v_T \int_{-\infty}^{d} y e^{-y^2/2} \, dy \right\} \\
&= e^{-rT} \left\{ (m_T - K) \Phi(d) + \frac{v_T}{\sqrt{2\pi}} e^{-d^2/2} \right\},
\end{aligned}$$

where $d = (m_T - K)/v_T$.

4.22

Note that by setting the risk-free interest rate equal to $\alpha - \sigma \lambda$ in the Black & Scholes formula in Thm. 4.6, we get

$$e^{-(\alpha - \sigma \lambda)T} \mathbb{E}_{\mathcal{Q}} \left[\max(S(T) - K, 0) \right] = S(0) \Phi(d_1) - K e^{-(\alpha - \sigma \lambda)T} \Phi(d_2),$$

with $d_1 = d_2 + \sigma \sqrt{T}$ and

$$d_2 = \frac{\ln(S(0)/K) + (\alpha - \sigma \lambda - \sigma^2/2)T}{\sigma \sqrt{T}}.$$

Hence, the price of a call on electricity is

$$\begin{aligned}
P(0) &= e^{-rT} \mathbb{E}_{\mathcal{Q}} \left[\max(S(T) - K, 0) \right] \\
&= e^{-rT} e^{(\alpha - \sigma \lambda)T} e^{-(\alpha - \sigma \lambda)T} \mathbb{E}_{\mathcal{Q}} \left[\max(S(T) - K, 0) \right] \\
&= e^{-(r - \alpha + \sigma \lambda)} S(0) \Phi(d_1) - K e^{-rT} \Phi(d_2).
\end{aligned}$$

5.1

We find the cumulative distribution of Y:

$$\mathcal{P}(Y \leq y) = \mathcal{P}\left(\Phi^{-1}(U) \leq y\right) = \mathcal{P}(U \leq \Phi(y)) = \Phi(y).$$

Hence, $Y \sim \mathcal{N}(0,1)$.

5.2

We find the partial derivatives of the function $u(\tau, y) = \exp(ay + b\tau)C(T - 2\tau/\sigma^2, e^y)$:

$$\frac{\partial u(\tau, y)}{\partial \tau} = bu(\tau, y) - \exp(ay + b\tau)\frac{\partial C(T - 2\tau/\sigma^2, e^y)}{\partial t}\frac{2}{\sigma^2},$$

$$\frac{\partial u(\tau, y)}{\partial y} = au(\tau, y) + \exp(ay + b\tau)\frac{\partial C(T - 2\tau/\sigma^2, e^y)}{\partial x}\exp(y),$$

and

$$\frac{\partial^2 u(\tau, y)}{\partial y^2} = a\frac{\partial u(\tau, y)}{\partial y} + a\exp(ay + b\tau)\frac{\partial C(T - 2\tau/\sigma^2, e^y)}{\partial x}\exp(y)$$

$$+ \exp(ay + b\tau)\frac{\partial^2 C(T - 2\tau/\sigma^2, e^y)}{\partial x^2}\exp(2y)$$

$$+ \exp(ay + b\tau)\frac{\partial C(T - 2\tau/\sigma^2, e^y)}{\partial x}\exp(y)$$

$$= (2a + 1)\frac{\partial u(\tau, y)}{\partial y} - (a^2 + a)u(\tau, y)$$

$$+ \exp(ay + b\tau)\frac{\partial^2 C(T - 2\tau/\sigma^2, e^y)}{\partial x^2}\exp(2y).$$

Since $C(t, x)$ is the solution of the Black & Scholes partial differential equation (4.12), we have

$$\frac{\partial C(t, x)}{\partial t} = -rx\frac{\partial C(t, x)}{\partial x} - \frac{1}{2}\sigma^2\frac{\partial^2 C(t, x)}{\partial x^2} + rC(t, x),$$

and thus setting $x = e^y$ and $t = T - 2\tau/\sigma^2$, yields

$$\frac{\partial u(\tau, y)}{\partial \tau} = bu(\tau, y) - \exp(ay + b\tau)\frac{\partial C(T - 2\tau/\sigma^2, e^y)}{\partial t}\frac{2}{\sigma^2}$$

$$= bu(\tau, y) + \frac{2r}{\sigma^2}\exp(ay + b\tau)\exp(y)\frac{\partial C(T - 2\tau/\sigma^2, e^y)}{\partial x}$$

$$+ \exp(2y)\exp(ay + b\tau)\frac{\partial^2 C(T - 2\tau/\sigma^2, e^y)}{\partial x^2}$$

$$- \frac{2r}{\sigma^2}\exp(ay + b\tau)C(T - 2\tau/\sigma^2, e^y)$$

$$= \left(b - \frac{2r(1 + a)}{\sigma^2} + a^2 + a\right)u(\tau, y) + \left(\frac{2r}{\sigma^2} - 2a - 1\right)\frac{\partial u(\tau, y)}{\partial y}$$

$$+ \frac{\partial^2 u(\tau, y)}{\partial y^2}.$$

By definition, $a = (2r/\sigma^2 - 1)/2$ and $b = (2r/\sigma^2 + 1)^2/4$. Therefore we have $2r/\sigma^2 = 2a + 1$ and $(1 + a)^2 = b$, which imply

$$b - \frac{2r(1+a)}{\sigma^2} + a^2 + a = (1+a)^2 - (2a+1)(1+a) + a^2 + a = 0,$$

and

$$\frac{2r}{\sigma^2} - 2a - 1 = 2a + 1 - 2a - 1 = 0.$$

Thus, we showed that $\partial u(\tau,y)/\partial \tau = \partial^2 u(\tau,y)/\partial y^2$.

5.3

From Subsect. 4.3.5 we have

$$C(t,x) = e^{-r(T-t)} \int_{-\infty}^{\infty} f(e^y) p(y;\sigma,r) \, dy,$$

with

$$p(y;\sigma,r) = \frac{1}{\sqrt{2\pi\sigma^2(T-t)}} \exp\left(-\frac{(y - \ln x - (r - \sigma^2/2)(T-t))^2}{2\sigma^2(T-t)}\right).$$

By differentiating the last expression, we find

$$\frac{\partial p(y;\sigma,r)}{\partial \sigma} = p(y;\sigma,r) \left\{ \frac{(y - \ln x - (r - \sigma^2/2)(T-t))^2}{\sigma^2(T-t)} - \frac{1}{\sigma} \right.$$
$$\left. - \frac{y - \ln x - (r - \sigma^2/2)(T-t)}{\sigma(T-t)} \right\}.$$

Hence, introducing the function

$$g(t,x,T,s) = f(s) \left\{ \frac{(\ln s - \ln x - (r - \sigma^2/2)(T-t))^2}{\sigma^2(T-t)} - \frac{1}{\sigma} \right.$$
$$\left. - \frac{\ln s - \ln x - (r - \sigma^2/2)(T-t)}{\sigma(T-t)} \right\},$$

we find

$$\frac{\partial C(t,x)}{\partial \sigma} = e^{-r(T-t)} \mathbb{E}_{\mathcal{Q}}\left[g(t,x,T,S^{t,x}(T))\right].$$

We can rewrite the vega in terms of an expectation of the increment $W(T) - W(t)$, where $W(t)$ is a Brownian motion with respect to $\mathcal{Q}$.

$$\frac{\partial C}{\partial \sigma}(t,x) = e^{-r(T-t)} \mathbb{E}_{\mathcal{Q}}\left[f\left(x \exp\left(\left(r - \frac{1}{2}\sigma^2\right)(T-t) + \sigma(W(T) - W(t))\right)\right)\right.$$
$$\left. \times \left\{\frac{(W(T) - W(t))^2}{\sigma^2(T-t)} - \frac{W(T) - W(t)}{\sigma(T-t)} - \frac{1}{\sigma}\right\}\right].$$

We can simulate the vega by appealing to Algorithm 1 with payoff function defined by $g(t,x,T,s)$, or we can create a tailor-made algorithm using the expectation of $W(T) - W(t)$. The second step in Algorithm 1 must then be modified accordingly.

The derivation of rho is left to the reader.

5.5

To simulate a *path* of a Brownian motion $B(t)$ requires us to draw an outcome of $B(t)$ for every t in an interval $[0,T]$. A computer can only simulate $B(t)$ for discrete times. Let $t_0 =< t_1 < t_2 < \ldots < t_{N-1} < t_N = T$ be a uniform partition of the interval $[0,T]$, with $t_j = j\Delta t$ and $\Delta t = T/N$, for $j = 0, \ldots, N$. We can then write $B(t)$ as

$$B(t_j) = B(t_{j-1}) + (B(t_j) - B(t_{j-1})).$$

The definition of Brownian motion says that $B(t_j) - B(t_{j-1})$ is a normally distributed random variable with mean zero and variance Δt. In addition, $B(t_j) - B(t_{j-1})$ is independent of $B(t_{j-1})$. Therefore,

$$B(t_j) = B(t_{j-1}) + \sqrt{\Delta t}\, Y_j,$$

where $Y_j := (B(t_j) - B(t_{j-1}))/\sqrt{\Delta t}$ is a standard normally distributed random variable which is independent of $B(t_{j-1})$, that is, Y_j is independent of Y_i for $i \neq j$. This gives us a recursive scheme to generate the path of a Brownian motion. An algorithm may look as follows:

Algorithm 6 *Algorithm for simulating the path of a Brownian motion.*

1. Draw N independent outcomes from the random variable $Y \sim \mathcal{N}(0,1)$:

$$(y^1, \ldots, y^N).$$

2. Set $b^0 = 0$.
3. For $j = 1, \ldots, N$, calculate

$$b^j = b^{j-1} + \sqrt{\Delta}\, y^j.$$

The numbers $b^0, b^1, \ldots, b^N$ will constitute one realization of the Brownian path at times $0, \Delta t, 2\Delta t, \ldots, T$. We end up with one path, being one outcome of the Brownian motion.[3]

Let us develop the analogous algorithm for geometric Brownian motion. We use the representation

$$S(t) = S(0)\exp\left(\mu t + \sigma B(t)\right).$$

By using the same partition as for Brownian motion above, we can write

---
[3] Note that when plotting the simulated path, one usually does a linear interpolation between the simulated outcomes.

$$S(t_j) = S(0)\exp\left(\mu t_j + \sigma B(t_j)\right)$$
$$= S(0)\exp\left(\mu t_{j-1} + \sigma B(t_{j-1})\right)\exp\left(\mu(t_j - t_{j-1}) + \sigma(B(t_j) - B(t_{j-1}))\right)$$
$$= S(t_{j-1})\exp\left(\mu \Delta t + \sigma\sqrt{\Delta t}\, Y_j\right).$$

We can use the following algorithm to simulate the path of a geometric Brownian motion:

Algorithm 7 *Algorithm for simulating the path of a geometric Brownian motion.*

1. Draw N independent outcomes from the random variable $Y \sim \mathcal{N}(0,1)$:
$$(y^1, \ldots, y^N).$$
2. Set $s^0 = S(0)$.
3. For $j = 1, \ldots, N$, calculate
$$s^j = s^{j-1} \exp\left(\mu \Delta t + \sigma \sqrt{\Delta t}\, y^j\right).$$

The numbers $s^0, \ldots, s^N$ will constitute one realization of the geometric Brownian motion at times $0, \Delta t, 2\Delta t, \ldots, T$.

References

1. Abramowitz, M. and Stegun, I.A.: Handbook of Mathematical Functions. Dover, New York (1972)
2. Bachelier, L.: Theory of speculation. Translation from the French in: P.H. Cootner (ed.), The Random Character of Stock Market Prices. The M.I.T. Press, Cambridge, Massachusetts (1964)
3. Barndorff-Nielsen, O.E.: Exponentially decreasing distributions for the logarithm of particle size. Proc. Roy. Soc. London Ser. A, **353**, 401–419 (1977)
4. Barndorff-Nielsen, O.E: Processes of normal inverse Gaussian type. Finance Stoch., **2**, 41–68 (1998)
5. Barndorff-Nielsen, O.E. and Shephard, N.: Financial Volatility: Stochastic Volatility and Lévy Based Models. Cambridge Unversity Press, Cambridge. *Monograph in preparation*
6. Benth, F.E. and Saltyte-Benth, J.: The normal inverse Gaussian distribution and spot price modelling in energy markets. E-print 26, Dept. Math., University of Oslo (2003). *To appear in* Int. J. Theor. Appl. Finance
7. Bjørk, Th.: Arbitrage Theory in Continuous Time. Oxford University Press, Oxford (1998)
8. Bølviken, E. and Benth, F.E.: Quantification of risk in Norwegian stocks via the normal inverse Gaussian distribution. In: R. Norberg et al. (eds) Proceedings of the 10th AFIR Colloquium in Tromsø 2000. AFIR, 87–98 (2000)
9. Black, F. and Scholes, M.: The pricing of options and corporate liabilities. J. Polit. Economy, **81**, 637–654 (1973)
10. Black, F.: The pricing of commodity contracts. J. Financ. Econom., **3**, 167–169 (1976)
11. Broadie, M. and Glasserman, P.: Estimating security price derivatives using simulation. Manag. Sci., **42**(2), 269–285 (1996)
12. Chan, T.: Pricing contingent claims on stocks driven by Lévy processes. Ann. Applied Probab., **9**, 504–528 (1999)
13. Clewlow, L. and Strickland, C.: Energy Derivatives – Pricing and Risk Management. Lacima Publications (2000)
14. Dahl, L.O.: An adaptive method for evaluating multidimensional contingent claims: Part I. Int. J. Theor. Appl. Finance, **6**(3), 301–316 (2003)
15. Dahl, L.O.: An adaptive method for evaluating multidimensional contingent claims: Part II. Int. J. Theor. Appl. Finance, **6**(4), 327–353 (2003)
16. Davis, M.H.A.: Pricing weather derivatives by marginal value. Quant. Finance, **1**, 305–308 (2001)

17. Davis, M.H.A., Panas, V.G. and Zariphopoulou, T.: European option pricing with transaction costs. SIAM J. Control Optim., **31**, 470–493 (1993)
18. Delbaen, F. and Schachermayer, W.: A general version of the fundamental theorem of asset pricing. Math. Ann., **300**, 463–520 (1994)
19. Doornik, J.A.: Ox: Object Oriented Matrix Programming 2.0. Timberlake, London (1998)
20. Duffie, D.: Dynamic Asset Pricing Theory (2nd Ed.). Princeton University Press, Princeton (1996)
21. Eberlein, E. and Jacod, J.: On the range of options prices. Finance Stoch., **1**, 131–140 (1997)
22. Eberlein, E. and Keller, U.: Hyperbolic distributions in finance. Bernoulli, **1**, 281–299 (1995)
23. Eberlein, E., Keller, U. and Prause, K.: New insights into smile, mispricing and Value at Risk: The hyperbolic model. J. Business, **71**, 371–405 (1998)
24. Eydeland, A. and Wolyniec, K.: Energy and Power Risk Management. John Wiley & Sons, New York, New Jersey (2003)
25. Folland, G.B.: Real Analysis. John Wiley & Sons, New York (1984)
26. Föllmer, H. and Schweizer, M.: Hedging of contingent claims under incomplete information. In: M.H.A. Davies and R.J. Elliot (eds) Applied Stochastic Analysis, 389–414, Gordon and Breach, New York (1991)
27. Fournie, E., Lasry, J.-M., Lebuchoux, J. Lions, P.-L. and Touzi, N.: Application of Malliavin calculus to Monte Carlo methods in finance. Finance Stoch., **3**, 391–412 (1999)
28. Gerber, H.U. and Shiu, E.S.W.: Option pricing by Esscher transforms. Trans. Soc. Actuaries, **46**, 99–191 (1994) (with discussion)
29. Hodges, S.D. and Neuberger, A.: Optimal replication of contingent claims under transaction costs. Rev. Futures Markets, **8**, 222–239 (1989)
30. Hull, J.C.: Options, Futures, and Other Derivative Securities. Prentice Hall, Englewood Cliffs (1993)
31. Ikeda, N. and Watanabe, S.: Stochastic Differential Equations and Diffusion Processes. North-Holland, Kodansha (1981)
32. Jäckel, P.: Monte Carlo Methods in Finance. John Wiley & Sons, Chichester (2002)
33. Karatzas, I. and Shreve, S.E.: Brownian Motion and Stochastic Calculus. Springer-Verlag, New York (1991)
34. Karatzas, I. and Shreve, S.E.: Methods of Mathematical Finance. Springer-Verlag, New York (1998)
35. Lehoczky, J.P.: Simulation methods for option pricing. In M. H. A. Dempster and S. R. Pliska (eds) Mathematics for Derivative Securities. Publication of Newton Institute, 529–543, Cambridge University Press, Cambridge (1997)
36. Mandelbrot, B.: The variation of certain speculative prices. J. Business, **36**, 394–419 (1963)
37. Mandelbrot, B.: Fractals and Scaling in Finance. Discontinuity, Concentration, Risk. Springer-Verlag, New York (1997)
38. Margrabe, W.: The value of an option to exchange one asset for another. J. Finance, **33**, 177–186 (1978)

39. Merton, R.: Theory of rational option pricing. Bell J. Econom. Manag. Sci., **4**, 141–183 (1973)
40. Musiela, M. and Rutkowski, M.: Martingale methods in financial modelling. Applications of Mathematics, 36. Springer-Verlag, Berlin (1997)
41. Øksendal, B.: Stochastic Differential Equations. An Introduction with Applications (5th Ed.). Springer-Verlag, Berlin Heidelberg (1998)
42. Øksendal, B.: An introduction to Malliavin calculus with applications to economics. Working paper No. 3/96, Norwegian School of Business and Administration, Bergen (1996). Can be downloaded from *http://www.nhh.no/for/dp/1996/wp0396.pdf*
43. Pitman, J.: Probability. Springer-Verlag, New York (1997).
44. Pliska, S.R.: Introduction to Mathematical Finance. Discrete Time Models. Blackwell Publishers, Massachusetts, Oxford (1997)
45. Prause, K.: The Generalized Hyperbolic Model: Estimation, Financial Derivatives, and Risk Measures. PhD thesis, University of Freiburg, Germany (1999)
46. Press, W.H., Teukolsky, S.A., Vetterling, W.T. and Flannery, B.P.: Numerical Recipes in C. Cambridge University Press, Cambridge (1992)
47. Protter, Ph.: Stochastic Integration and Differential Equations. A New Approach. Springer-Verlag, Berlin Heidelberg (1990)
48. Rydberg, T.H.: The normal inverse Gaussian Lévy process: Simulation and approximation. Commun. Statist.–Stoch. Models, **13**, 887–910 (1997)
49. Schwartz, E.S.: The stochastic behaviour of commodity prices: Implications for valuation and hedging. J. Finance, **LII**(3), 923–973 (1997)

Index

arbitrage   4, 56, 59
autocorrelation   29

Bessel function   24
Black & Scholes
– formula   66
– – implied volatility   72
– – volatility smile   73
– market   58
Brownian motion   12
– path   13, 119

Cauchy–Schwarz inequality   50
central limit theorem   9
complete market   56, 59, 87
contingent claim   53
– path dependent   107
correlation   8
covariance   8

density   *see* probability density
derivatives   *see* option
distribution
– generalized hyperbolic   23
– normal inverse Gaussian   24
– – shape triangle   26
distribution function   7

equivalent martingale measure   55, 74
– multi-dimensional   85
estimation   9
expectation   7
– conditional   8, 46

finite difference method   6, 113
forward contract   96
fundamental theorem of asset pricing
   86

geometric Brownian motion   12, 45
– drift   12
– multi-dimensional   45, 83
– path   119
– stochastic volatility   88
– volatility   12
Girsanov's theorem   75
– multi-dimensional   84

hedging   61
– delta-   62
– density approach   70

incomplete market   57, 88
independence   8
Itô integral   36
Itô isometry   37
Itô's formula   38
– for Brownian motion   40
– general   41
– multi-dimensional   43

Lévy process   12
– normal inverse Gaussian   28
lognormal process   12
logreturn   16

Malliavin derivative   79, 112
market risk of price   93
martingale   *see* stochastic process
martingale representation theorem   49
maximum likelihood   9
mean   7
moment generating function   51
Monte Carlo   6, 99
– low discrepancy sequence   107
– quasi   107

numerical differentiation   103

Index

option 2
- Asian 54, 110
- average 54, 110
- barrier 53
- basket 83
- Bermudan 110
- call 1, 66, 95
- chooser 79
- delta 62, 102
- digital 95
- electricity spot call 93
- hedge 4
- knock-out 53, 108
- put 1, 95
- quadratic 119
- rho 119
- spread 83, 97
- vega 119

portfolio 58
- doubling strategy 60
- self-financing 59
- sub-replicating 94
- super-replicating 94
probability 6
probability density 7
- joint 8
probability space 6
put–call parity 80, 95

quantile 7

random variable 7
- Gaussian 7
- lognormal 7
- normal 7
replicating strategy *see* hedging
risk-neutral probability *see* equivalent martingale measure

semimartingale *see* stochastic process
standard deviation 8
state space 7
stochastic differential equation 45
stochastic process 12
- adapted 35
- Itô integrability 36
- Markov property 82
- martingale 46
- semimartingale 41
stochastic volatility 88

value-at-risk 23, 32
VaR *see* value-at-risk
variance 7
volatility 12, 66, 84
- implied 72
- smile 73

Universitext

Aksoy, A.; Khamsi, M. A.: Methods in Fixed Point Theory

Alevras, D.; Padberg M. W.: Linear Optimization and Extensions

Andersson, M.: Topics in Complex Analysis

Aoki, M.: State Space Modeling of Time Series

Audin, M.: Geometry

Aupetit, B.: A Primer on Spectral Theory

Bachem, A.; Kern, W.: Linear Programming Duality

Bachmann, G.; Narici, L.; Beckenstein, E.: Fourier and Wavelet Analysis

Badescu, L.: Algebraic Surfaces

Balakrishnan, R.; Ranganathan, K.: A Textbook of Graph Theory

Balser, W.: Formal Power Series and Linear Systems of Meromorphic Ordinary Differential Equations

Bapat, R.B.: Linear Algebra and Linear Models

Benedetti, R.; Petronio, C.: Lectures on Hyperbolic Geometry

Benth, F. E.: Option Theory with Stochastic Analysis

Berberian, S. K.: Fundamentals of Real Analysis

Berger, M.: Geometry I, and II

Bliedtner, J.; Hansen, W.: Potential Theory

Blowey, J. F.; Coleman, J. P.; Craig, A. W. (Eds.): Theory and Numerics of Differential Equations

Börger, E.; Grädel, E.; Gurevich, Y.: The Classical Decision Problem

Böttcher, A; Silbermann, B.: Introduction to Large Truncated Toeplitz Matrices

Boltyanski, V.; Martini, H.; Soltan, P. S.: Excursions into Combinatorial Geometry

Boltyanskii, V. G.; Efremovich, V. A.: Intuitive Combinatorial Topology

Booss, B.; Bleecker, D. D.: Topology and Analysis

Borkar, V. S.: Probability Theory

Carleson, L.; Gamelin, T. W.: Complex Dynamics

Cecil, T. E.: Lie Sphere Geometry: With Applications of Submanifolds

Chae, S. B.: Lebesgue Integration

Chandrasekharan, K.: Classical Fourier Transform

Charlap, L. S.: Bieberbach Groups and Flat Manifolds

Chern, S.: Complex Manifolds without Potential Theory

Chorin, A. J.; Marsden, J. E.: Mathematical Introduction to Fluid Mechanics

Cohn, H.: A Classical Invitation to Algebraic Numbers and Class Fields

Curtis, M. L.: Abstract Linear Algebra

Curtis, M. L.: Matrix Groups

Cyganowski, S.; Kloeden, P.; Ombach, J.: From Elementary Probability to Stochastic Differential Equations with MAPLE

Dalen, D. van: Logic and Structure

Das, A.: The Special Theory of Relativity: A Mathematical Exposition

Debarre, O.: Higher-Dimensional Algebraic Geometry

Deitmar, A.: A First Course in Harmonic Analysis

Demazure, M.: Bifurcations and Catastrophes

Devlin, K. J.: Fundamentals of Contemporary Set Theory

DiBenedetto, E.: Degenerate Parabolic Equations

Diener, F.; Diener, M.(Eds.): Nonstandard Analysis in Practice

Dimca, A.: Singularities and Topology of Hypersurfaces

DoCarmo, M. P.: Differential Forms and Applications

Duistermaat, J. J.; Kolk, J. A. C.: Lie Groups

Edwards, R. E.: A Formal Background to Higher Mathematics Ia, and Ib

Edwards, R. E.: A Formal Background to Higher Mathematics IIa, and IIb

Emery, M.: Stochastic Calculus in Manifolds

Endler, O.: Valuation Theory

Erez, B.: Galois Modules in Arithmetic

Everest, G.; Ward, T.: Heights of Polynomials and Entropy in Algebraic Dynamics

Farenick, D. R.: Algebras of Linear Transformations

Foulds, L. R.: Graph Theory Applications

Frauenthal, J. C.: Mathematical Modeling in Epidemiology

Friedman, R.: Algebraic Surfaces and Holomorphic Vector Bundles

Fuks, D. B.; Rokhlin, V. A.: Beginner's Course in Topology

Fuhrmann, P. A.: A Polynomial Approach to Linear Algebra

Gallot, S.; Hulin, D.; Lafontaine, J.: Riemannian Geometry

Gardiner, C. F.: A First Course in Group Theory

Gårding, L.; Tambour, T.: Algebra for Computer Science

Godbillon, C.: Dynamical Systems on Surfaces

Godement, R.: Analysis I

Goldblatt, R.: Orthogonality and Spacetime Geometry

Gouvêa, F. Q.: p-Adic Numbers

Gustafson, K. E.; Rao, D. K. M.: Numerical Range. The Field of Values of Linear Operators and Matrices

Gustafson, S. J.; Sigal, I. M.: Mathematical Concepts of Quantum Mechanics

Hahn, A. J.: Quadratic Algebras, Clifford Algebras, and Arithmetic Witt Groups

Hájek, P.; Havránek, T.: Mechanizing Hypothesis Formation

Heinonen, J.: Lectures on Analysis on Metric Spaces

Hlawka, E.; Schoißengeier, J.; Taschner, R.: Geometric and Analytic Number Theory

Holmgren, R. A.: A First Course in Discrete Dynamical Systems

Howe, R., Tan, E. Ch.: Non-Abelian Harmonic Analysis

Howes, N. R.: Modern Analysis and Topology

Hsieh, P.-F.; Sibuya, Y. (Eds.): Basic Theory of Ordinary Differential Equations

Humi, M., Miller, W.: Second Course in Ordinary Differential Equations for Scientists and Engineers

Hurwitz, A.; Kritikos, N.: Lectures on Number Theory

Iversen, B.: Cohomology of Sheaves

Jacod, J.; Protter, P.: Probability Essentials

Jennings, G. A.: Modern Geometry with Applications

Jones, A.; Morris, S. A.; Pearson, K. R.: Abstract Algebra and Famous Inpossibilities

Jost, J.: Compact Riemann Surfaces

Jost, J.: Postmodern Analysis

Jost, J.: Riemannian Geometry and Geometric Analysis

Kac, V.; Cheung, P.: Quantum Calculus

Kannan, R.; Krueger, C. K.: Advanced Analysis on the Real Line

Kelly, P.; Matthews, G.: The Non-Euclidean Hyperbolic Plane

Kempf, G.: Complex Abelian Varieties and Theta Functions

Kitchens, B. P.: Symbolic Dynamics

Kloeden, P.; Ombach, J.; Cyganowski, S.: From Elementary Probability to Stochastic Differential Equations with MAPLE

Kloeden, P. E.; Platen; E.; Schurz, H.: Numerical Solution of SDE Through Computer Experiments

Kostrikin, A. I.: Introduction to Algebra

Krasnoselskii, M. A.; Pokrovskii, A. V.: Systems with Hysteresis

Luecking, D. H., Rubel, L. A.: Complex Analysis. A Functional Analysis Approach

Ma, Zhi-Ming; Roeckner, M.: Introduction to the Theory of (non-symmetric) Dirichlet Forms

Mac Lane, S.; Moerdijk, I.: Sheaves in Geometry and Logic

Marcus, D. A.: Number Fields

Martinez, A.: An Introduction to Semiclassical and Microlocal Analysis

Matoušek, J.: Using the Borsuk-Ulam Theorem

Matsuki, K.: Introduction to the Mori Program

Mc Carthy, P. J.: Introduction to Arithmetical Functions

Meyer, R. M.: Essential Mathematics for Applied Field

Meyer-Nieberg, P.: Banach Lattices

Mikosch, T.: Non-Life Insurance Mathematics

Mines, R.; Richman, F.; Ruitenburg, W.: A Course in Constructive Algebra

Moise, E. E.: Introductory Problem Courses in Analysis and Topology

Montesinos-Amilibia, J. M.: Classical Tessellations and Three Manifolds

Morris, P.: Introduction to Game Theory

Nikulin, V. V.; Shafarevich, I. R.: Geometries and Groups

Oden, J. J.; Reddy, J. N.: Variational Methods in Theoretical Mechanics

Øksendal, B.: Stochastic Differential Equations

Poizat, B.: A Course in Model Theory

Polster, B.: A Geometrical Picture Book

Porter, J. R.; Woods, R. G.: Extensions and Absolutes of Hausdorff Spaces

Radjavi, H.; Rosenthal, P.: Simultaneous Triangularization

Ramsay, A.; Richtmeyer, R. D.: Introduction to Hyperbolic Geometry

Rees, E. G.: Notes on Geometry

Reisel, R. B.: Elementary Theory of Metric Spaces

Rey, W. J. J.: Introduction to Robust and Quasi-Robust Statistical Methods

Ribenboim, P.: Classical Theory of Algebraic Numbers

Rickart, C. E.: Natural Function Algebras

Rotman, J. J.: Galois Theory

Rubel, L. A.: Entire and Meromorphic Functions

Rybakowski, K. P.: The Homotopy Index and Partial Differential Equations

Sagan, H.: Space-Filling Curves

Samelson, H.: Notes on Lie Algebras

Schiff, J. L.: Normal Families

Sengupta, J. K.: Optimal Decisions under Uncertainty

Séroul, R.: Programming for Mathematicians

Seydel, R.: Tools for Computational Finance

Shafarevich, I. R.: Discourses on Algebra

Shapiro, J. H.: Composition Operators and Classical Function Theory

Simonnet, M.: Measures and Probabilities

Smith, K. E.; Kahanpää, L.; Kekäläinen, P.; Traves, W.: An Invitation to Algebraic Geometry

Smith, K. T.: Power Series from a Computational Point of View

Smoryński, C.: Logical Number Theory I. An Introduction

Stichtenoth, H.: Algebraic Function Fields and Codes

Stillwell, J.: Geometry of Surfaces

Stroock, D. W.: An Introduction to the Theory of Large Deviations

Sunder, V. S.: An Invitation to von Neumann Algebras

Tamme, G.: Introduction to Étale Cohomology

Tondeur, P.: Foliations on Riemannian Manifolds

Verhulst, F.: Nonlinear Differential Equations and Dynamical Systems

Wong, M. W.: Weyl Transforms

Xambó-Descamps, S.: Block Error-Correcting Codes

Zaanen, A.C.: Continuity, Integration and Fourier Theory

Zhang, F.: Matrix Theory

Zong, C.: Sphere Packings

Zong, C.: Strange Phenomena in Convex and Discrete Geometry

Zorich, V. A.: Mathematical Analysis I

Zorich, V. A.: Mathematical Analysis II

Druck und Bindung: Strauss Offsetdruck GmbH